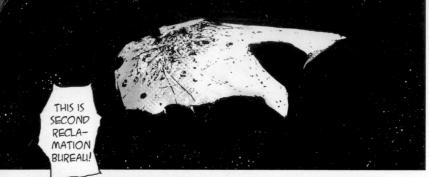

THIS IS SECOND RECLAMATION BUREAU!

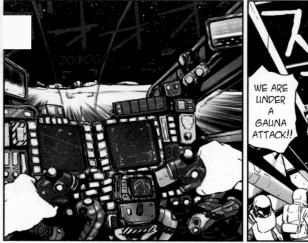

ZOOOOM

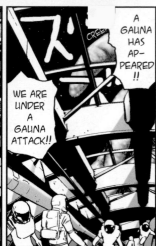

CREEE

A GAUNA HAS APPEARED!!

WE ARE UNDER A GAUNA ATTACK!!

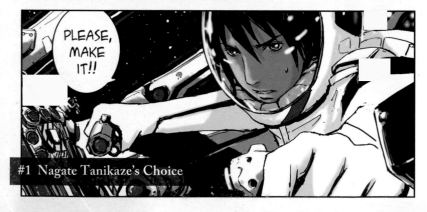

PLEASE, MAKE IT!!

#1 Nagate Tanikaze's Choice

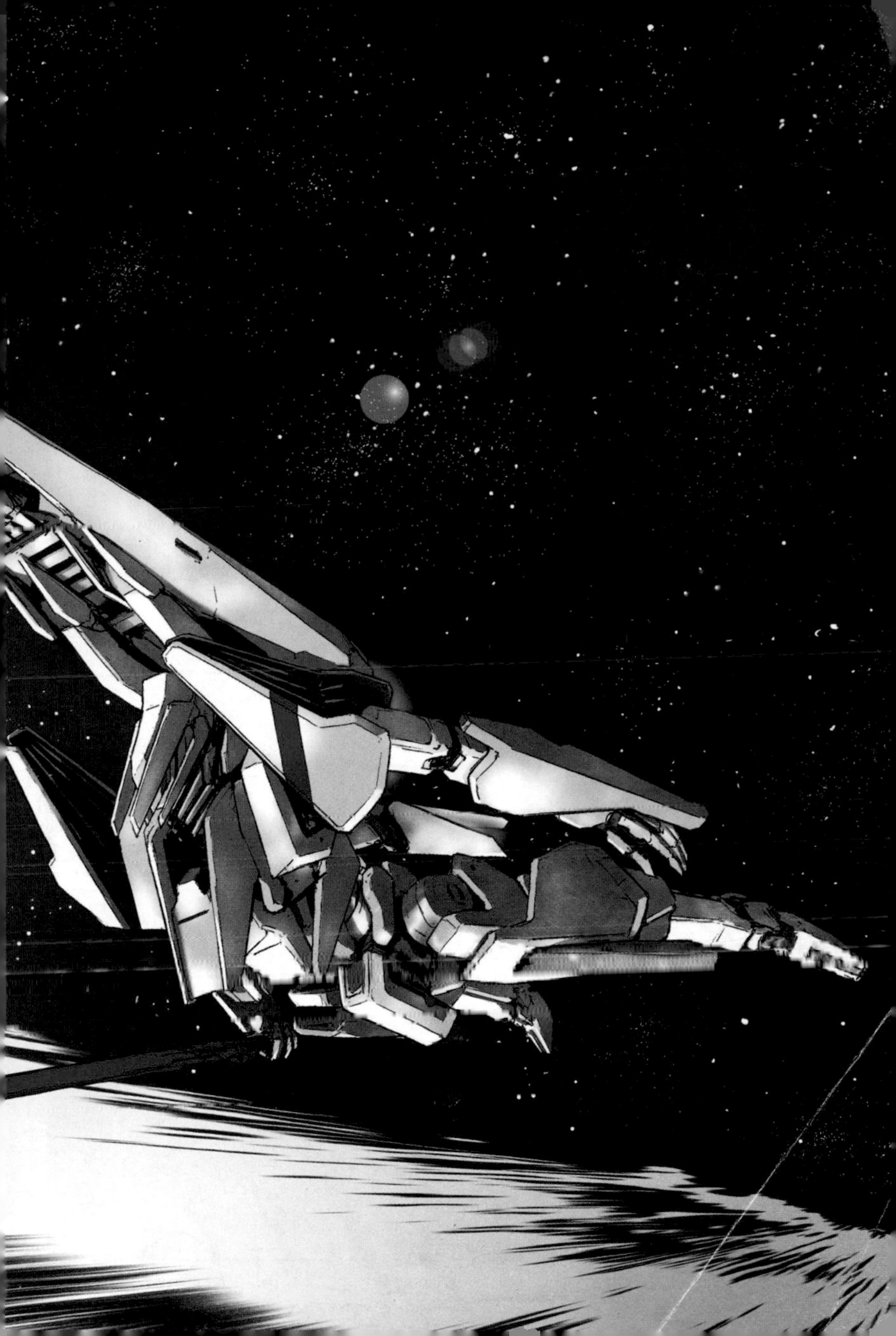

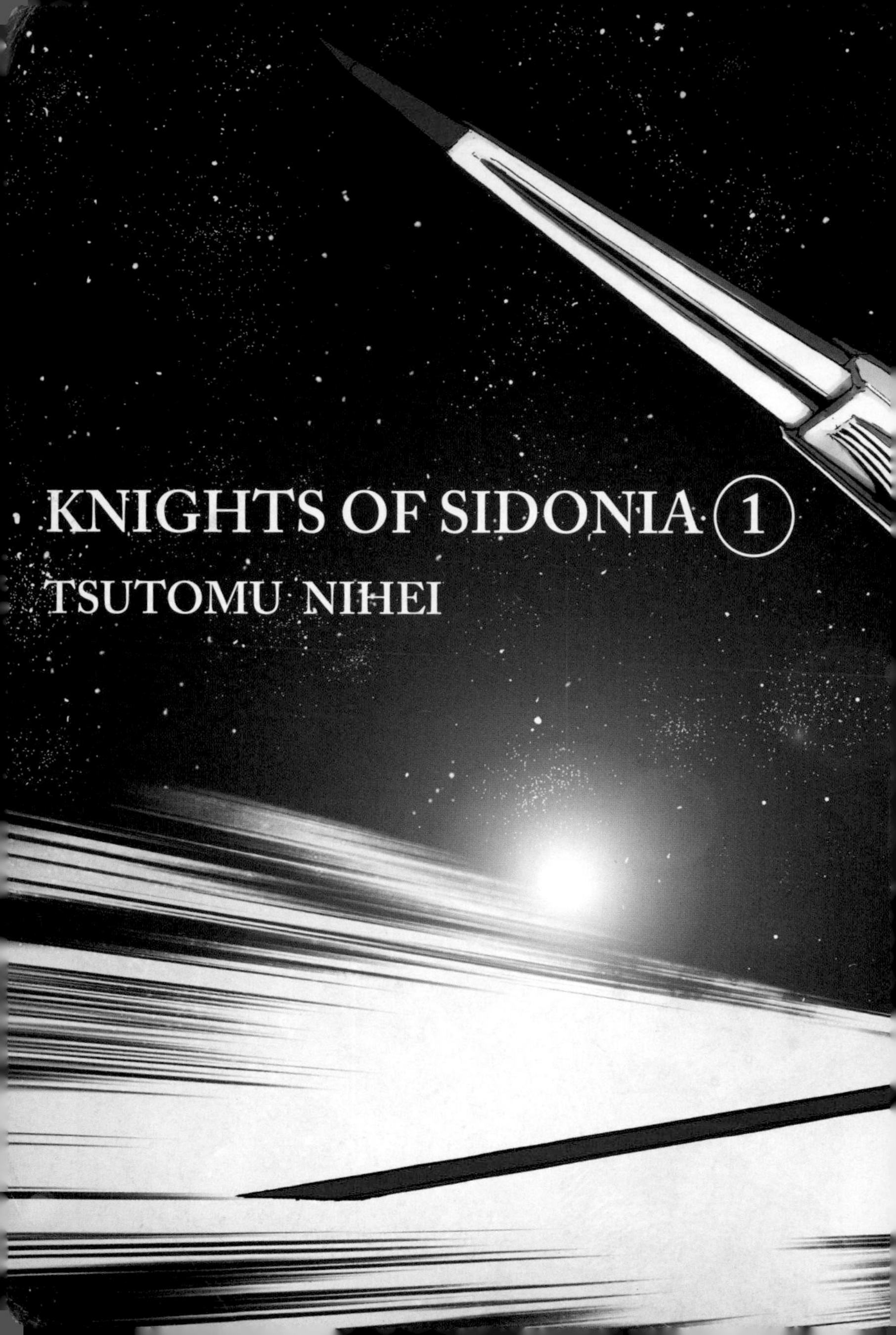
KNIGHTS OF SIDONIA 1
TSUTOMU NIHEI

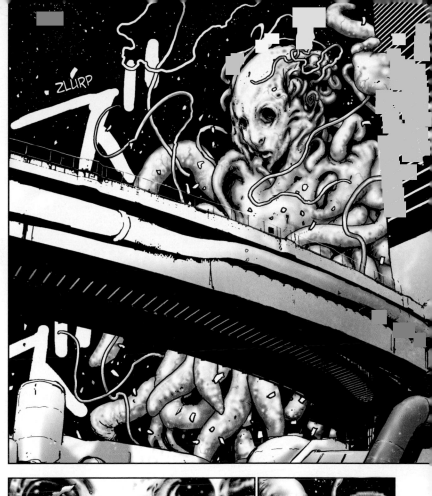

ZLURP

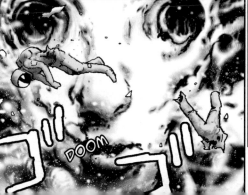

DOOM

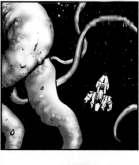

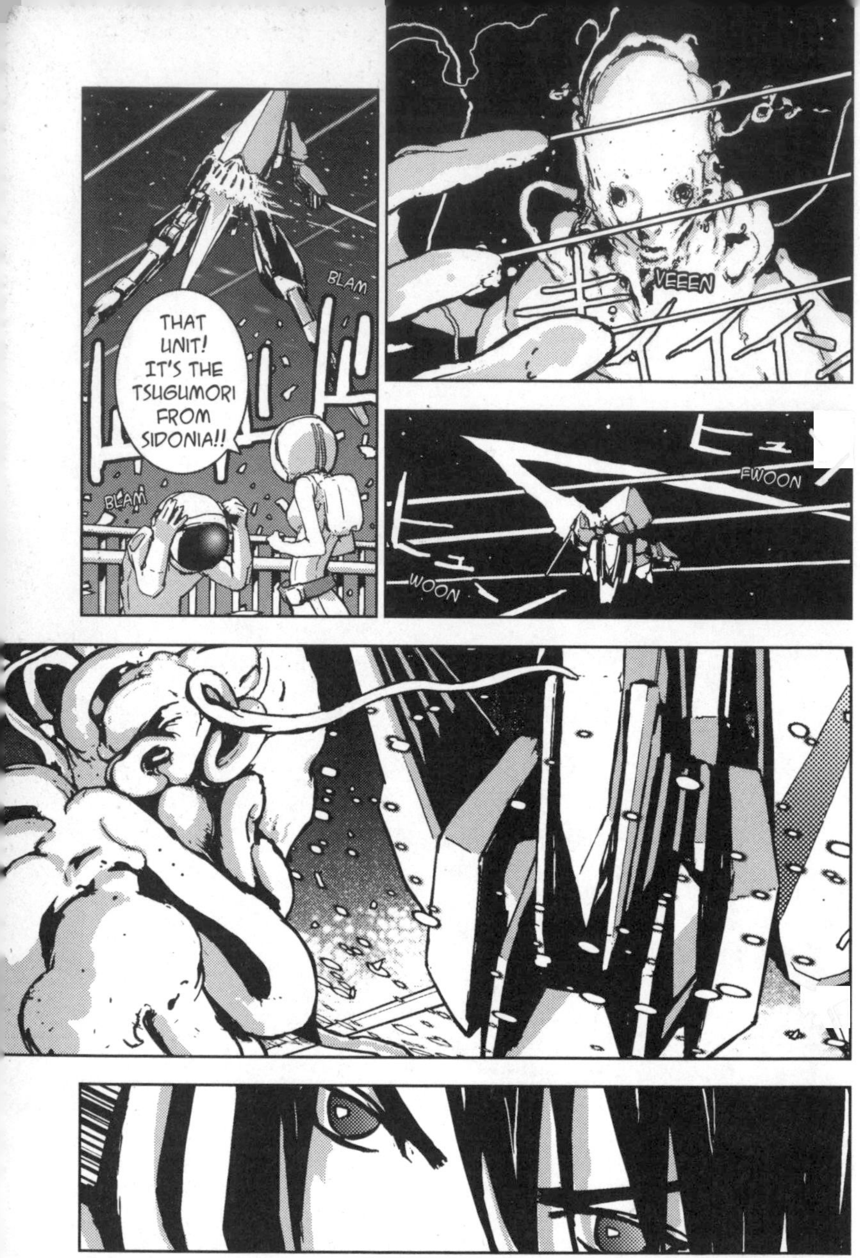
VEEEN
FWOON
WOON
BLAM
THAT UNIT! IT'S THE TSUGUMORI FROM SIDONIA!!
BLAM

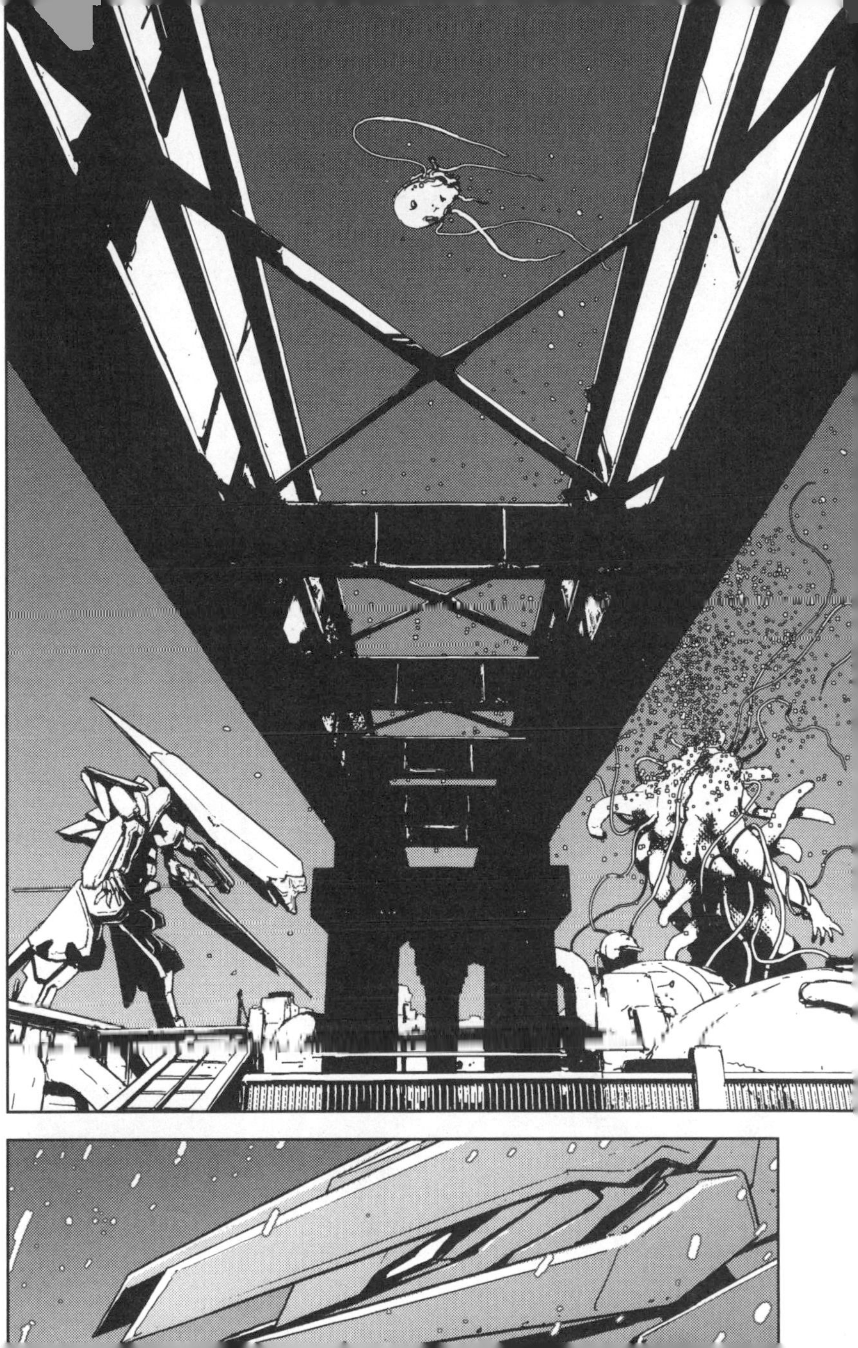

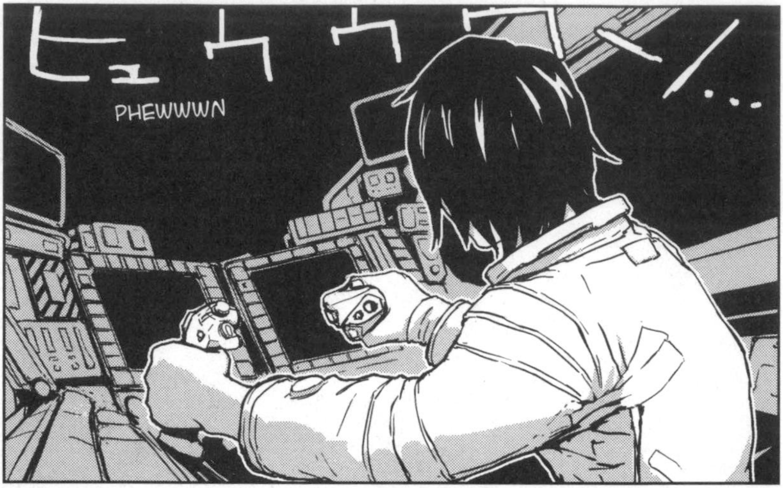
ヒュウウウン…
PHEWWWN

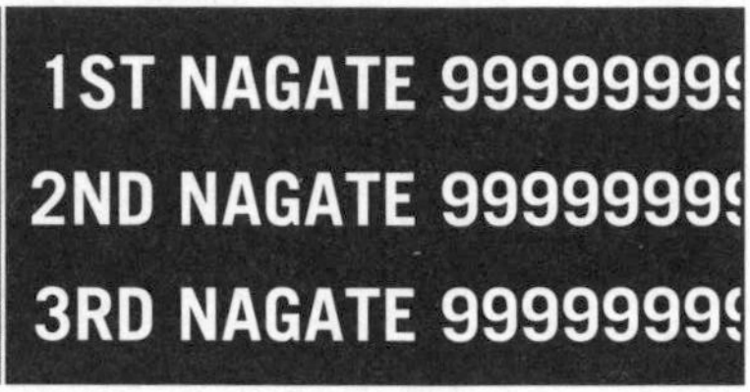
1ST NAGATE 99999999
2ND NAGATE 99999999
3RD NAGATE 99999999

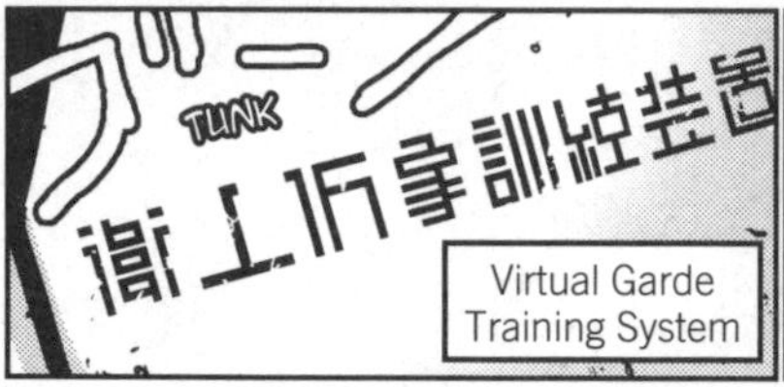
ブーン
TUNK
Virtual Garde Training System

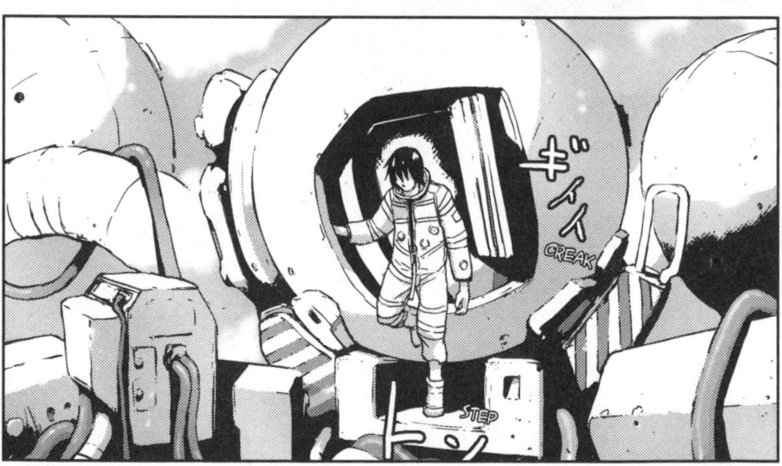
ギィイ
CREAK
トン
STEP

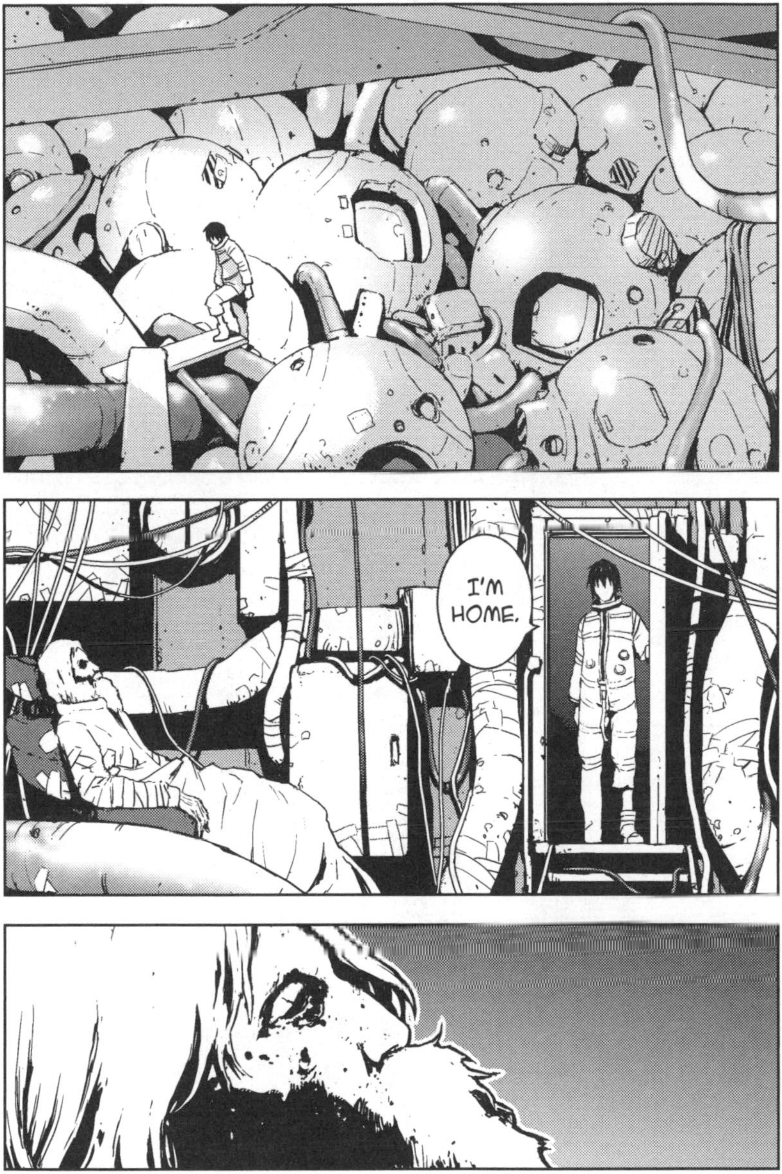
I'M HOME.

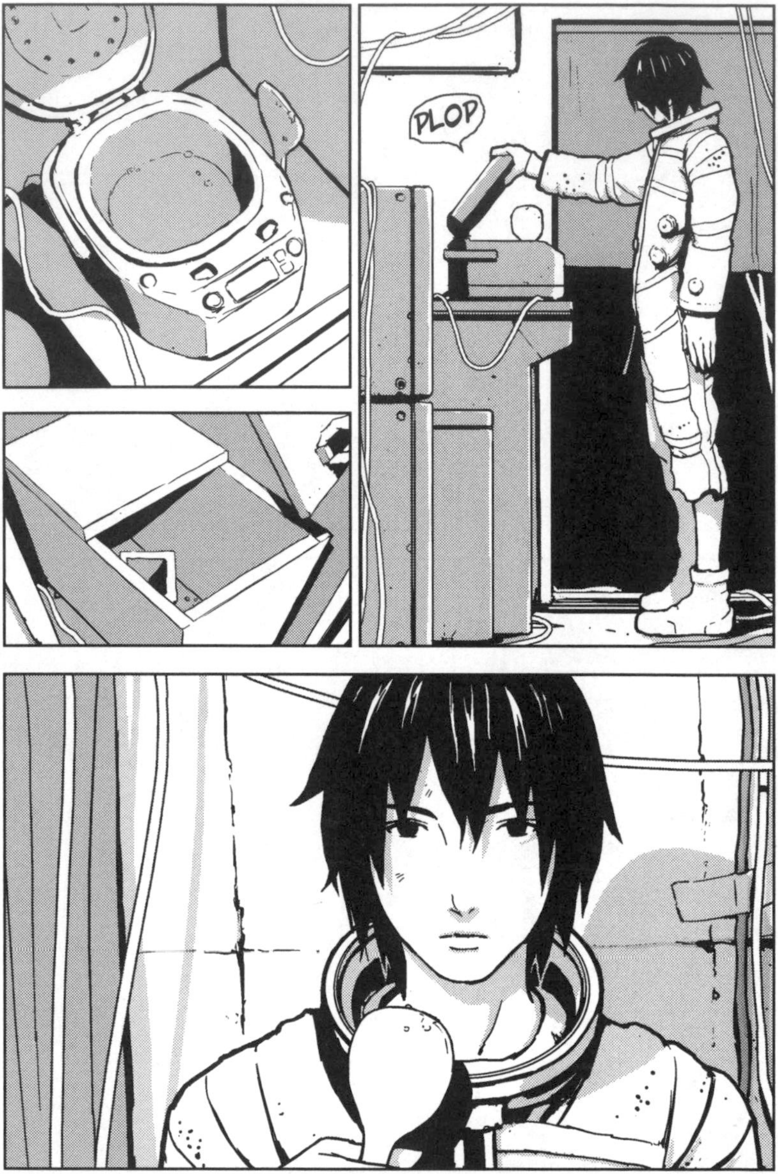
PLOP

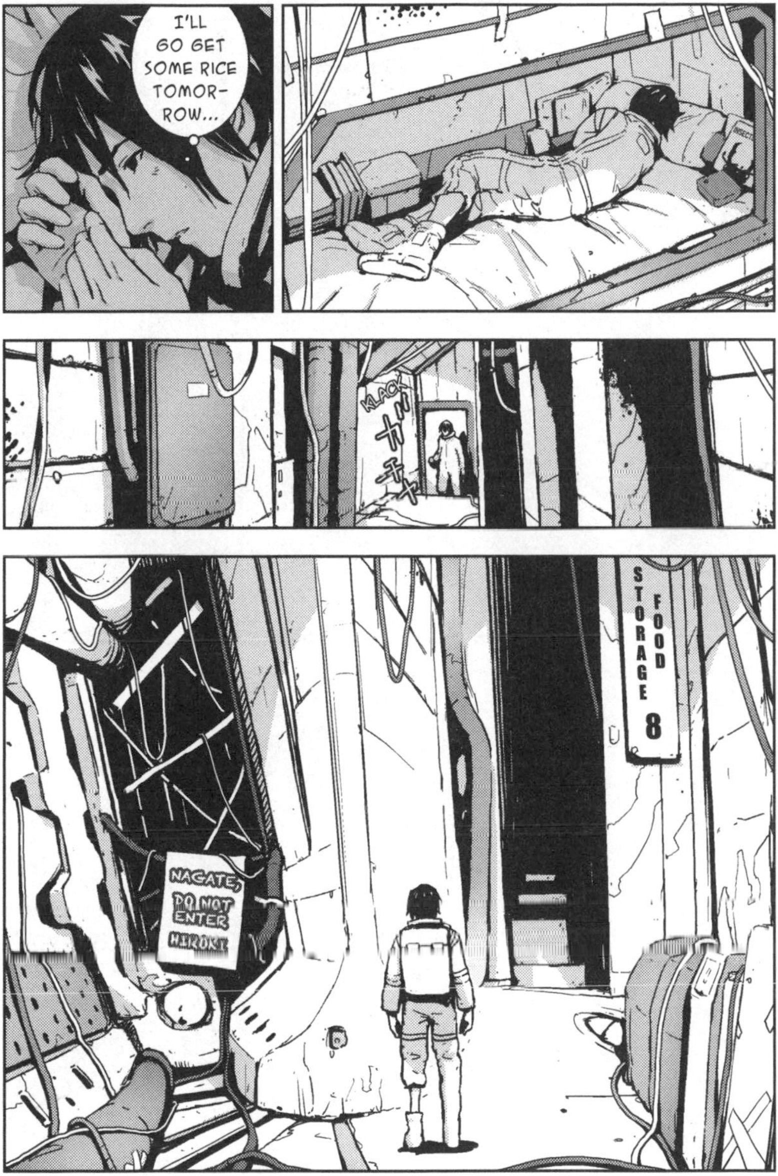
I'LL GO GET SOME RICE TOMORROW...
KLACK
ガチャ
FOOD STORAGE
8
NAGATE, DO NOT ENTER
HIROKI

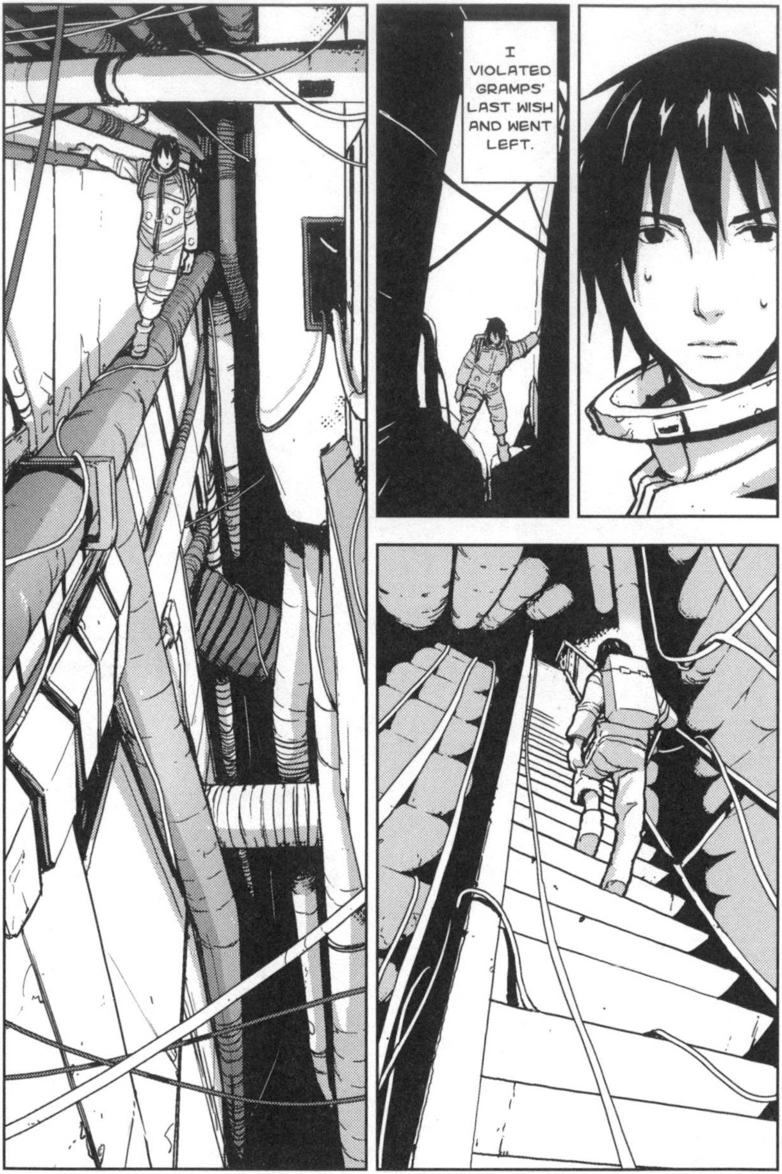
I VIOLATED GRAMPS' LAST WISH AND WENT LEFT.

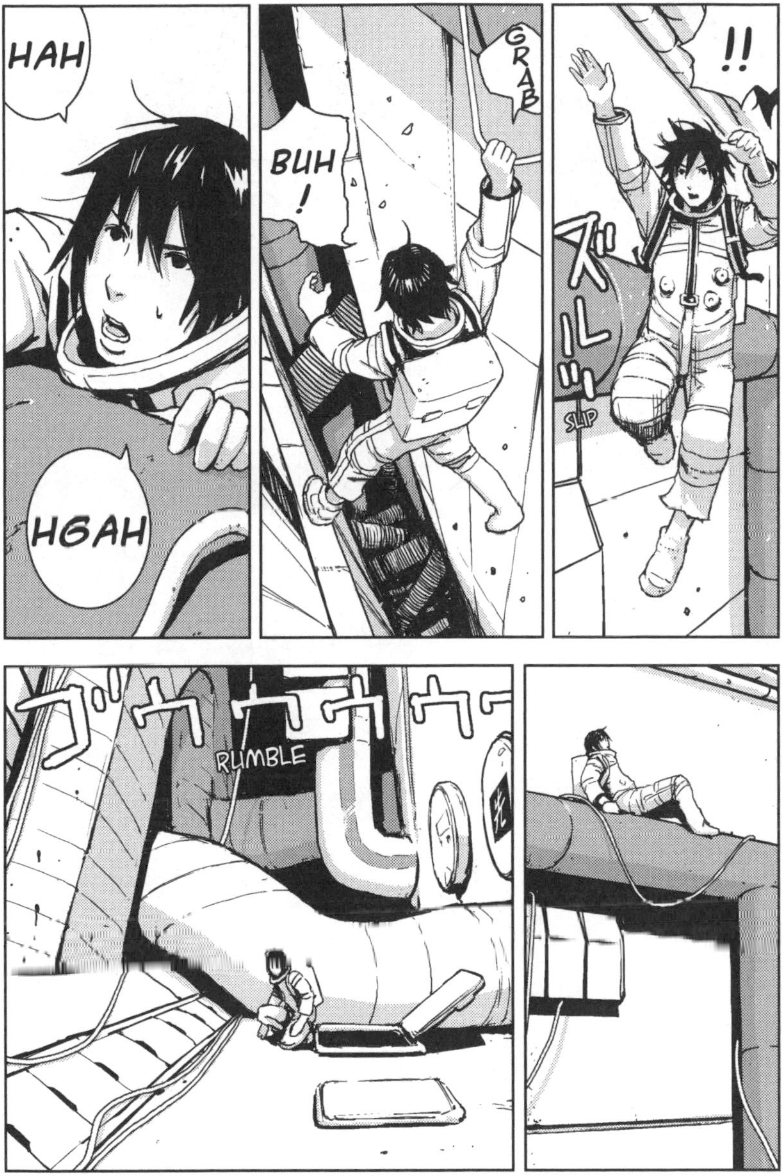
!!
ズルッ
SLIP
GRAB
BUH!
HAH
HGAH
ゴウウウウー
RUMBLE

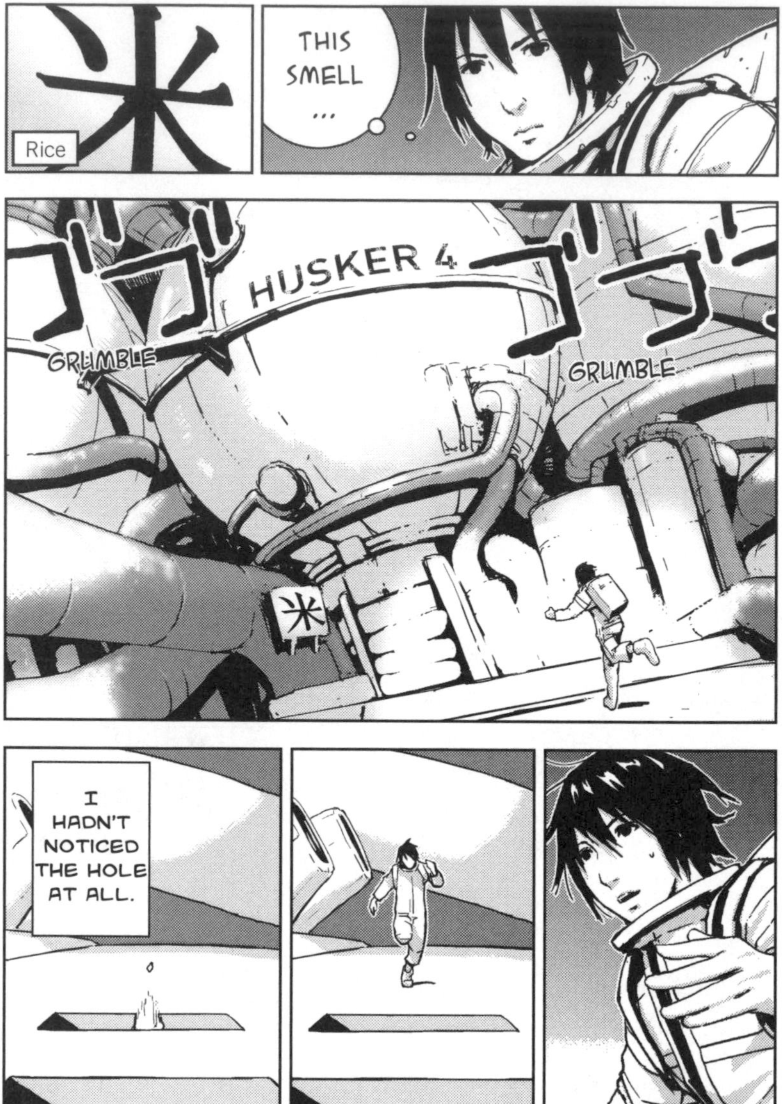
THIS SMELL ...
米
Rice
HUSKER 4
GRUMBLE
GRUMBLE
米
I HADN'T NOTICED THE HOLE AT ALL.

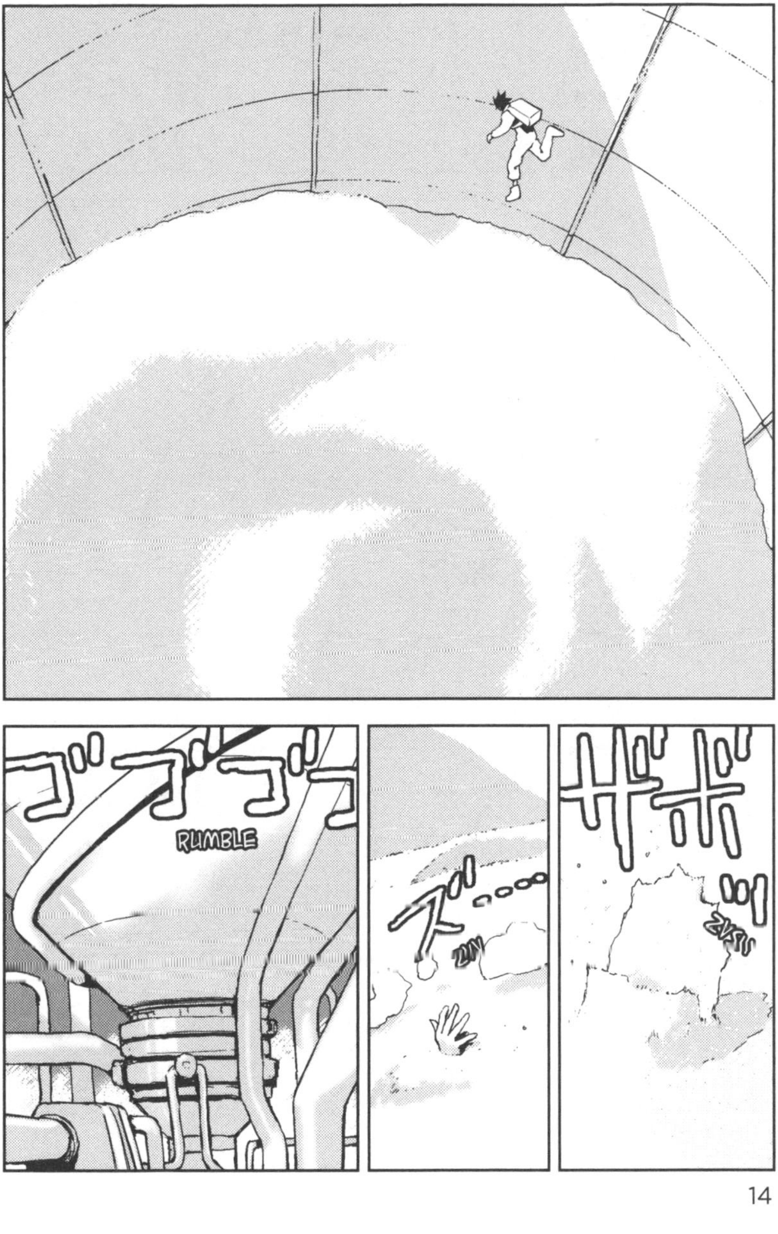
ザボッ
ズ゛ーーーー
ZUN
ゴゴゴゴ
RUMBLE

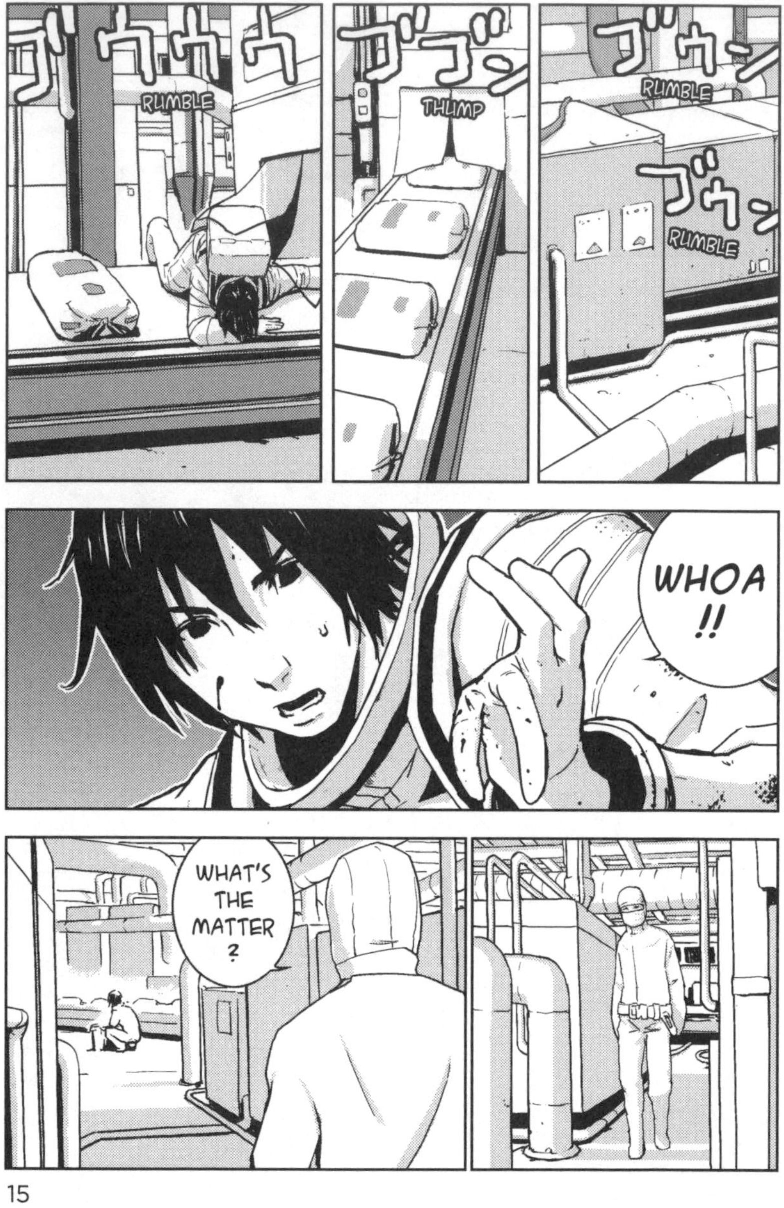
ゴウーン
RUMBLE
ゴウン
RUMBLE
ガゴン
THUMP
ゴウウウウ
RUMBLE
WHOA !!
WHAT'S THE MATTER ?

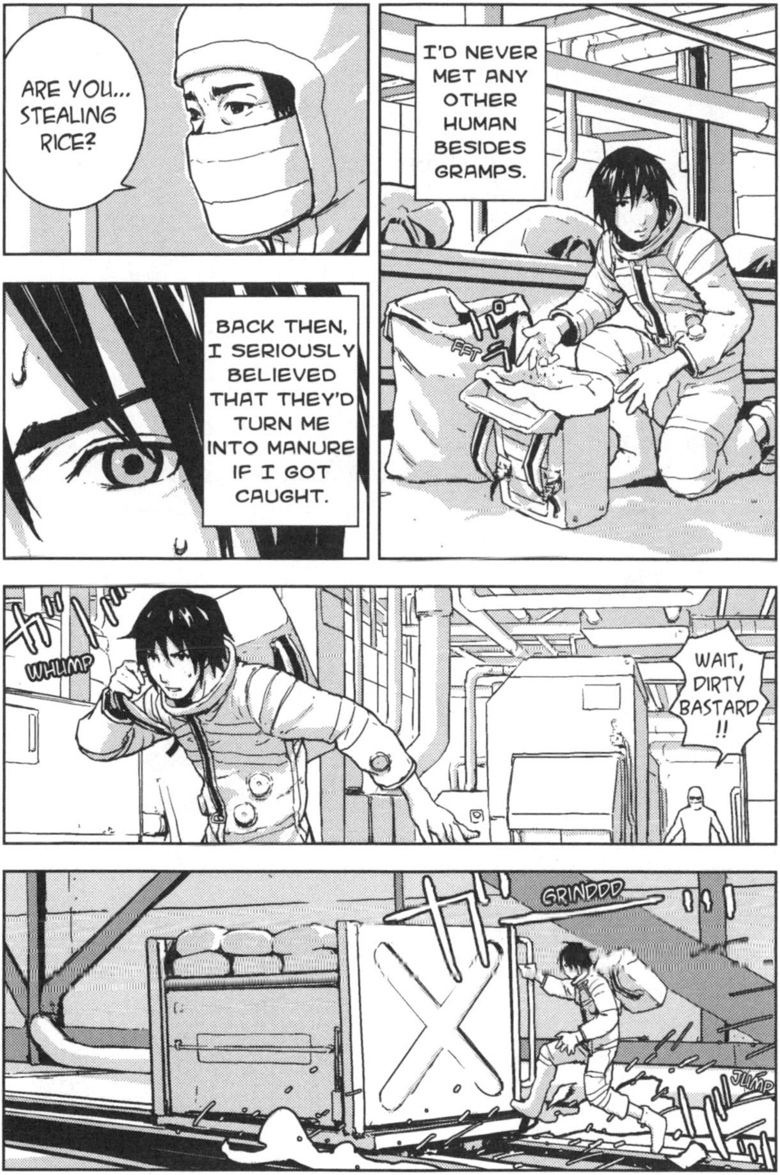
I'D NEVER MET ANY OTHER HUMAN BESIDES GRAMPS.
FFST
ARE YOU... STEALING RICE?
BACK THEN, I SERIOUSLY BELIEVED THAT THEY'D TURN ME INTO MANURE IF I GOT CAUGHT.
WAIT, DIRTY BASTARD !!
WHUMP
GRINDDD
JUMP

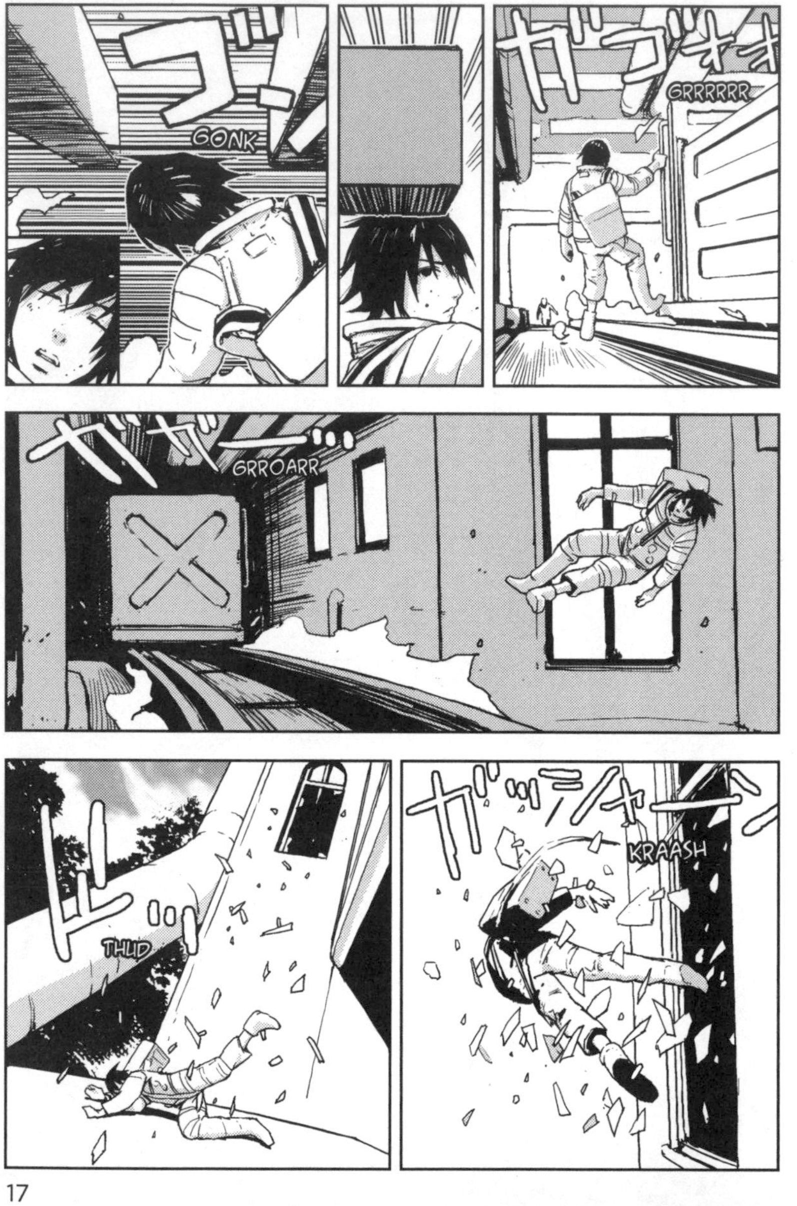
GRRRRRR
GONK
GRROARR
KRAASH
THUD

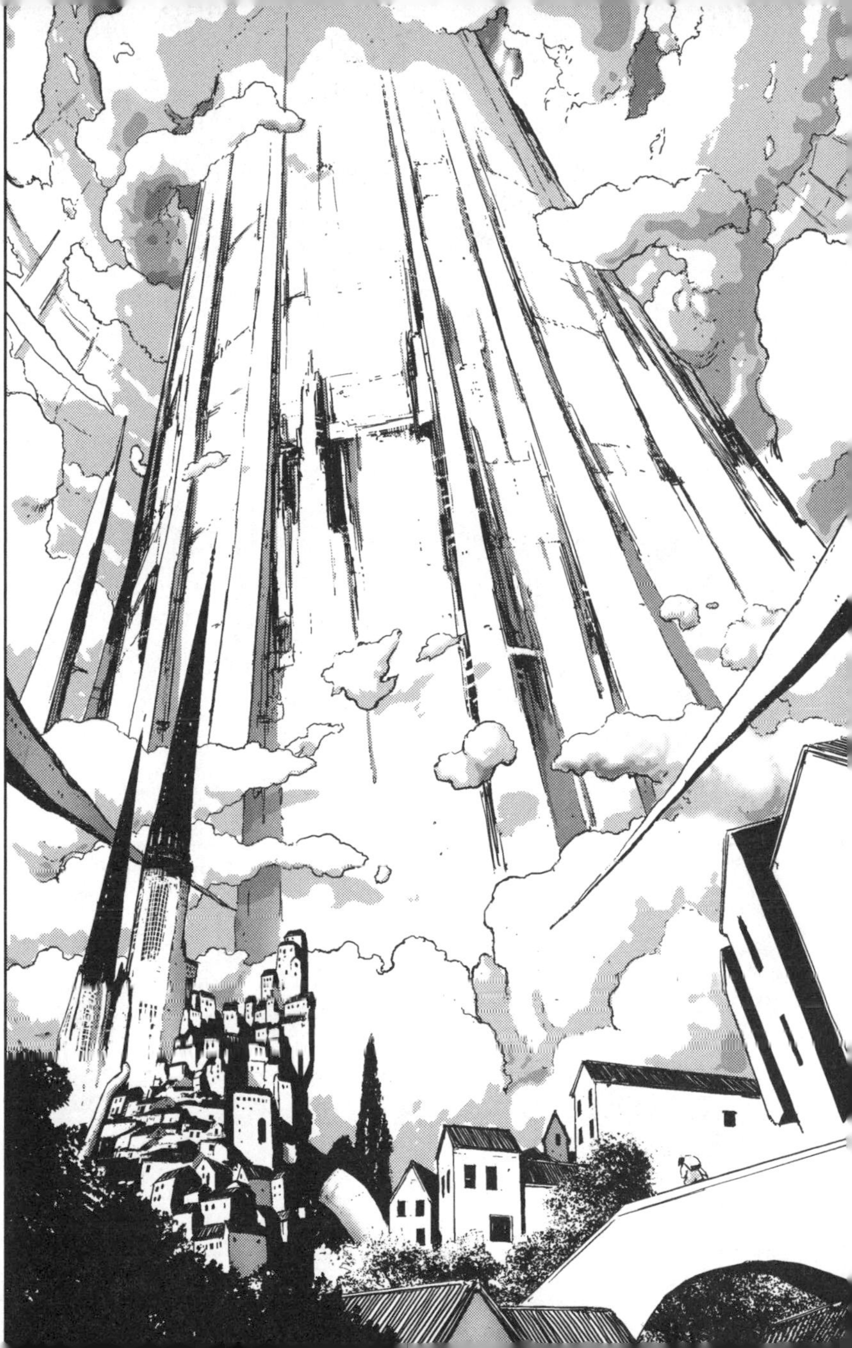

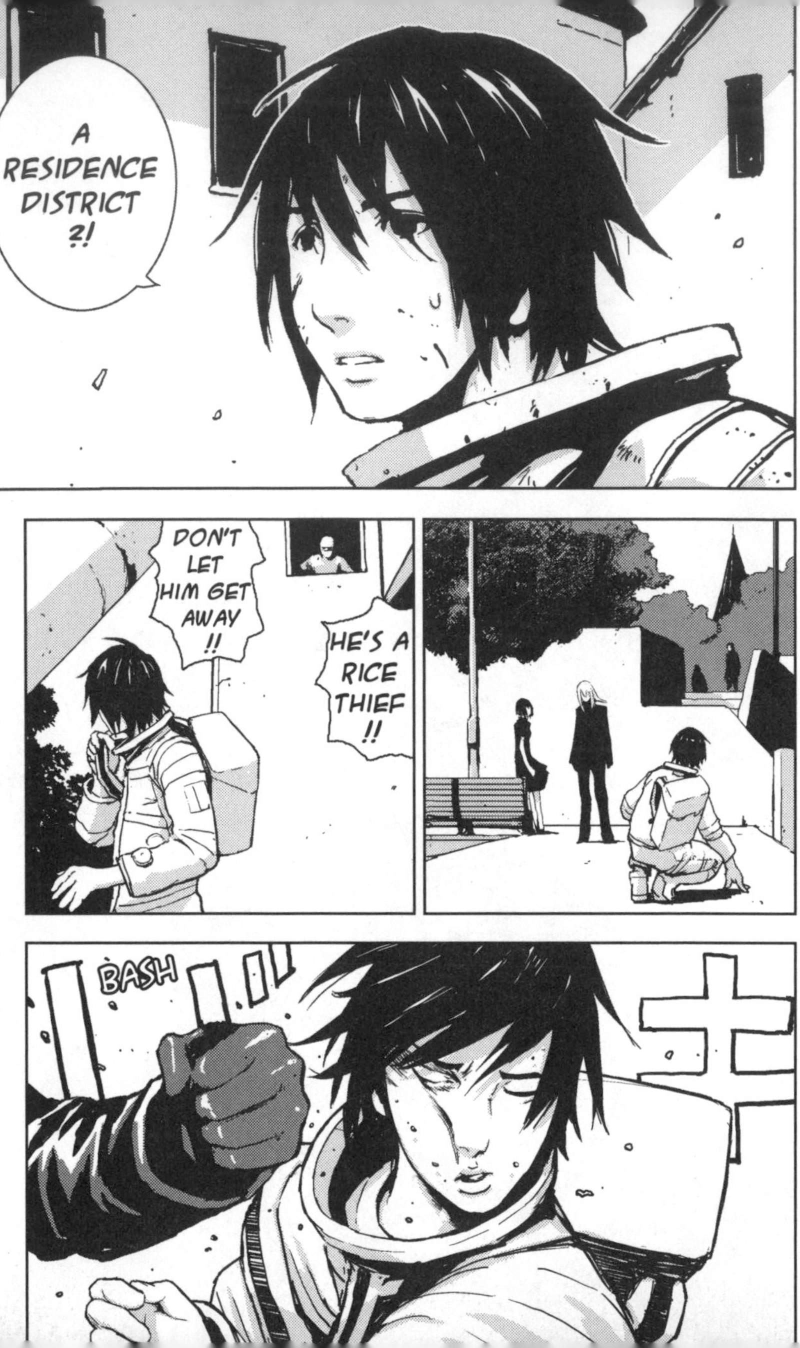
A RESIDENCE DISTRICT ?!
DON'T LET HIM GET AWAY !!
HE'S A RICE THIEF !!
BASH

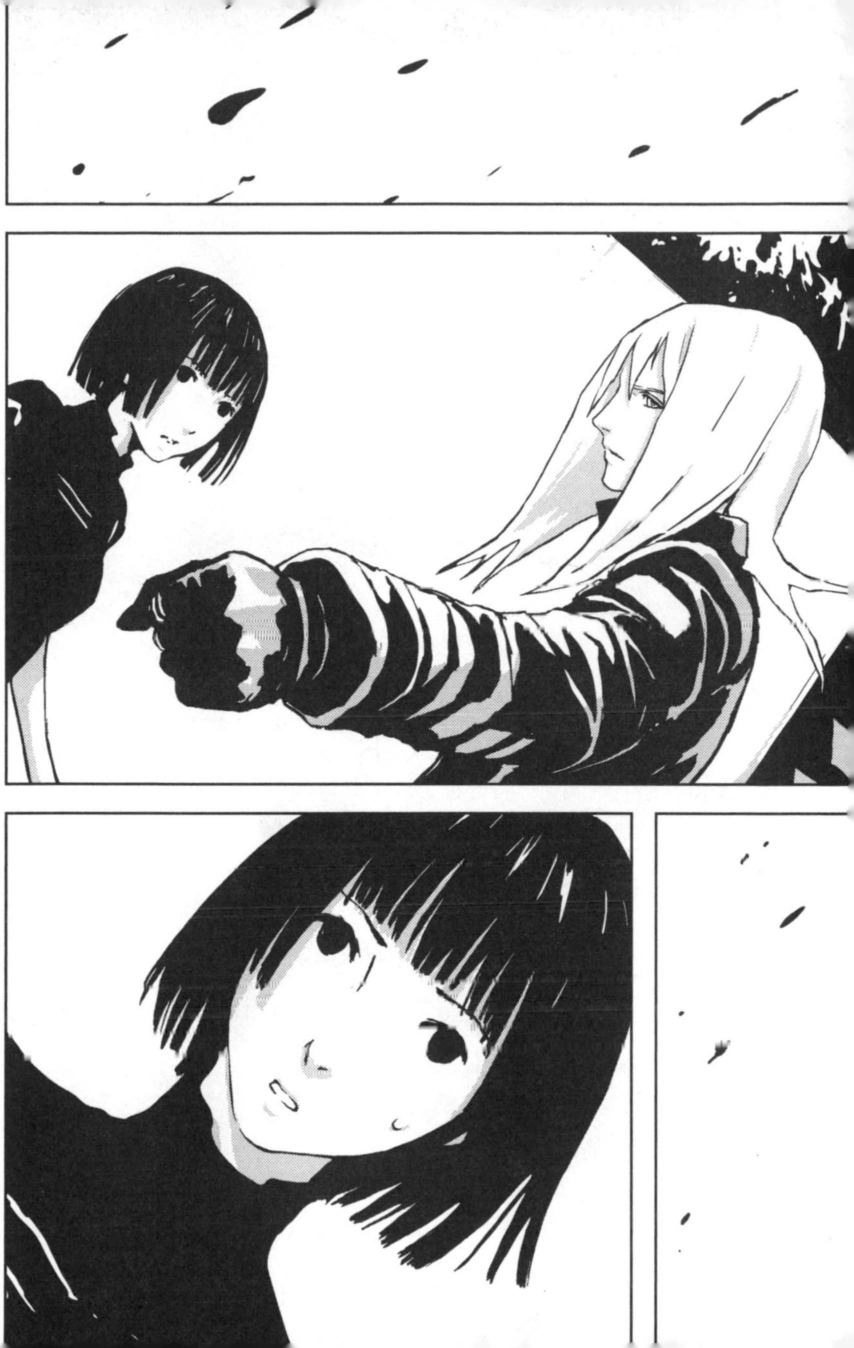

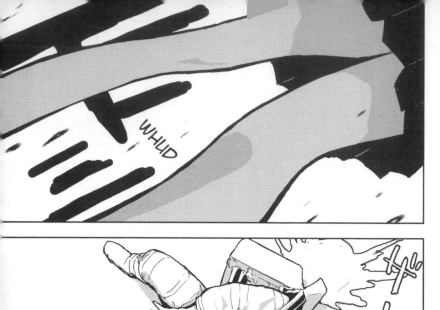

WHUD

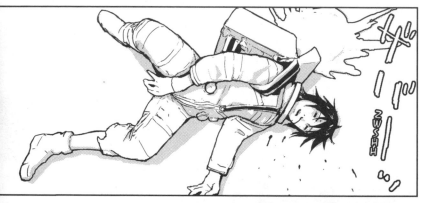

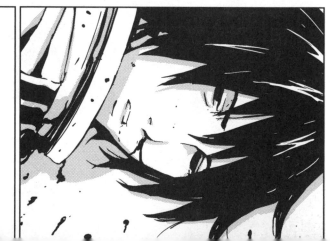

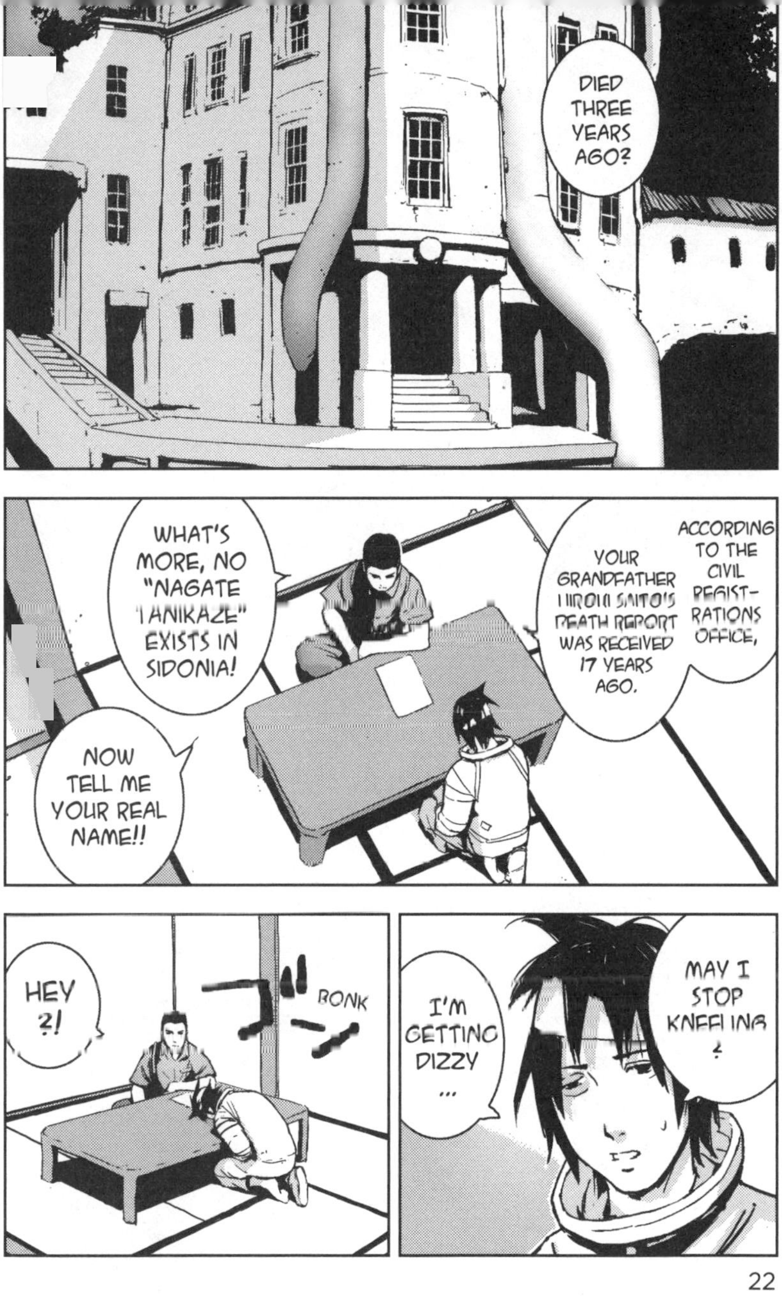
DIED THREE YEARS AGO?
ACCORDING TO THE CIVIL REGIST-RATIONS OFFICE,
YOUR GRANDFATHER HIROKI SAITO'S DEATH REPORT WAS RECEIVED 17 YEARS AGO.
WHAT'S MORE, NO "NAGATE TANIKAZE" EXISTS IN SIDONIA!
NOW TELL ME YOUR REAL NAME!!
MAY I STOP KNEELING ?
I'M GETTING DIZZY ...
BONK
HEY ?!

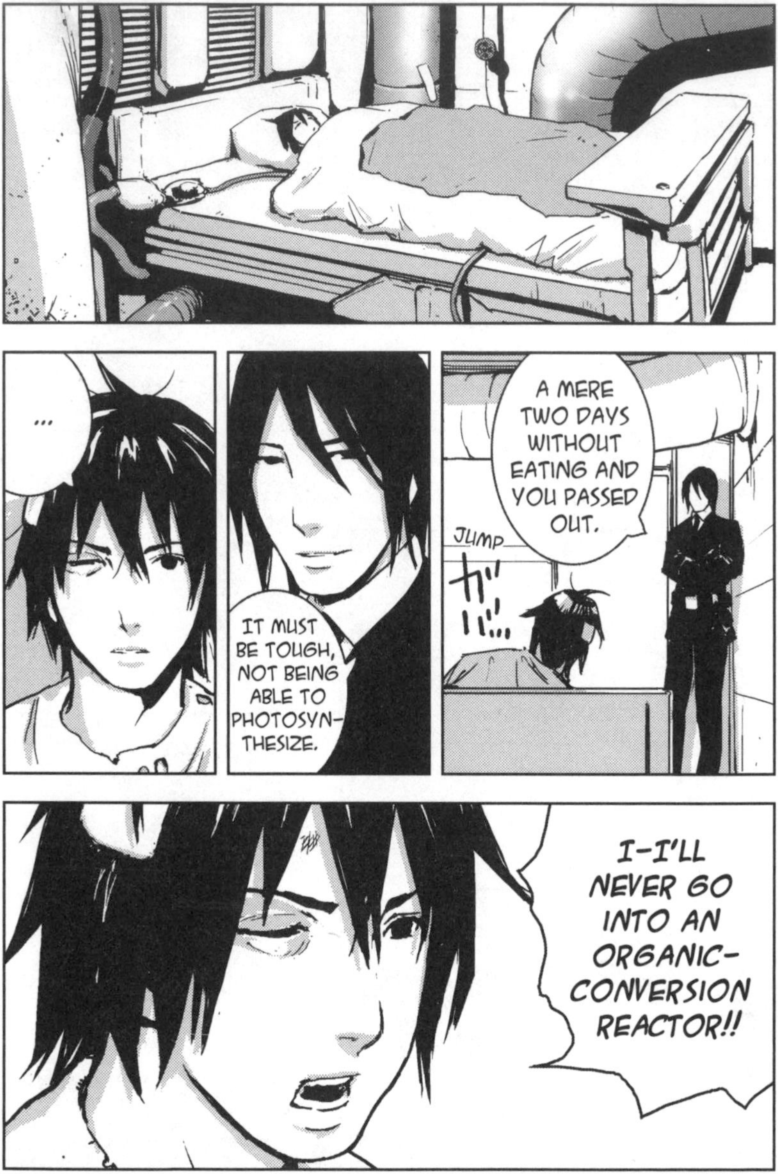
A MERE TWO DAYS WITHOUT EATING AND YOU PASSED OUT.
JUMP
IT MUST BE TOUGH, NOT BEING ABLE TO PHOTOSYN-THESIZE.
...
I-I'LL NEVER GO INTO AN ORGANIC-CONVERSION REACTOR!!

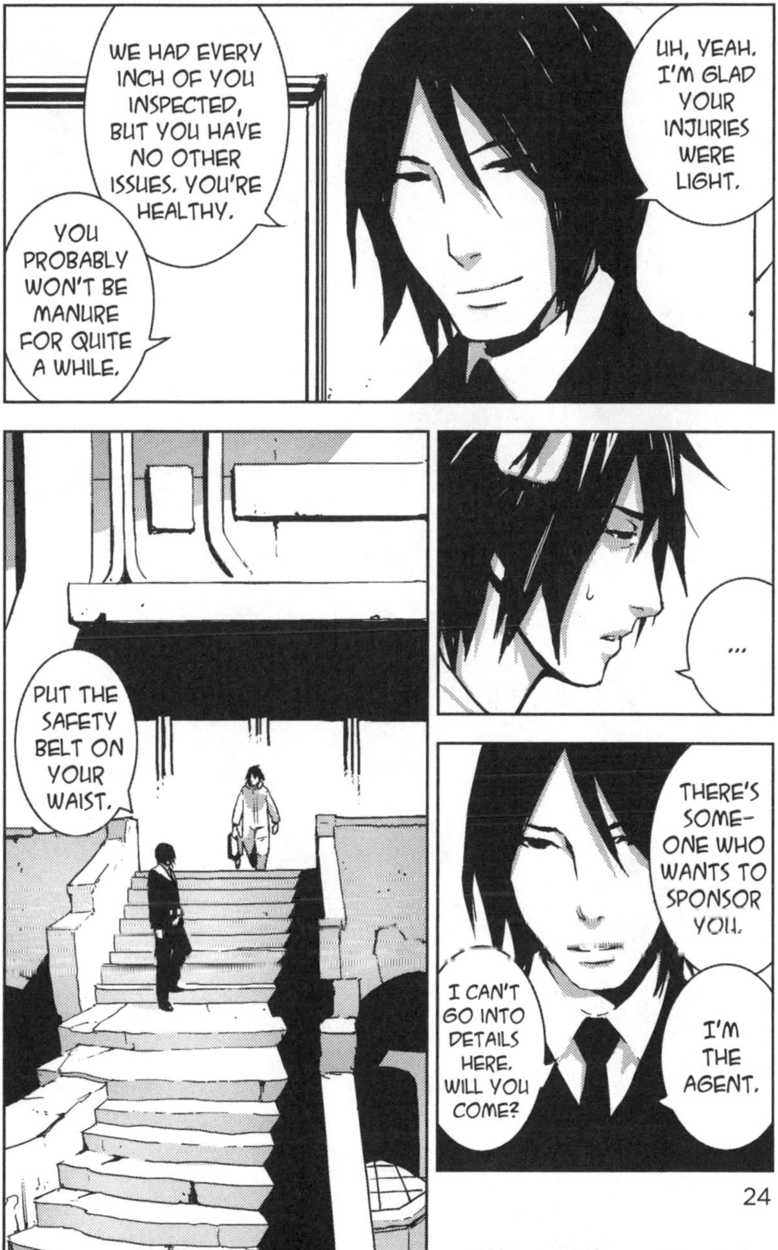
UH, YEAH. I'M GLAD YOUR INJURIES WERE LIGHT.
WE HAD EVERY INCH OF YOU INSPECTED, BUT YOU HAVE NO OTHER ISSUES. YOU'RE HEALTHY.
YOU PROBABLY WON'T BE MANURE FOR QUITE A WHILE.
...
THERE'S SOME-ONE WHO WANTS TO SPONSOR YOU.
I'M THE AGENT.
I CAN'T GO INTO DETAILS HERE. WILL YOU COME?
PUT THE SAFETY BELT ON YOUR WAIST.

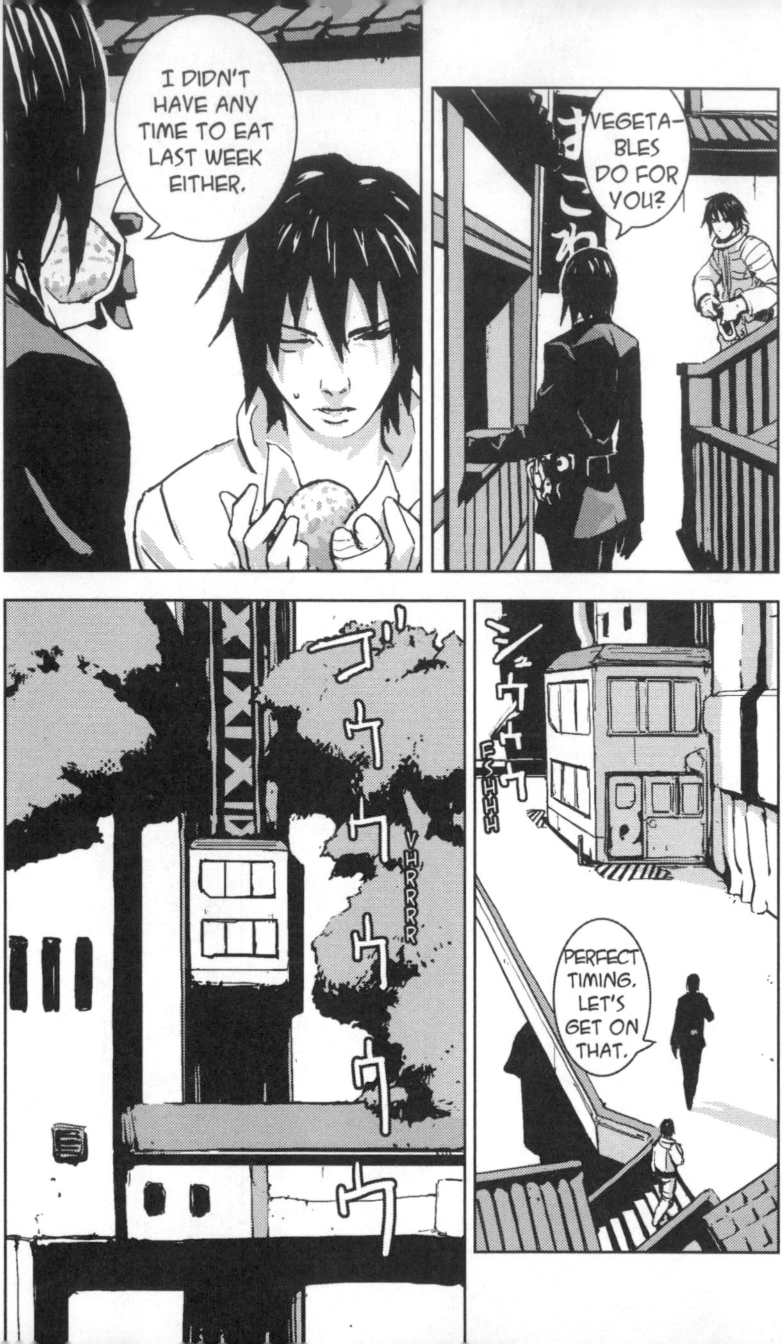
VEGETA-BLES DO FOR YOU?
I DIDN'T HAVE ANY TIME TO EAT LAST WEEK EITHER.
SHHHH
PERFECT TIMING. LET'S GET ON THAT.
VHRRRR

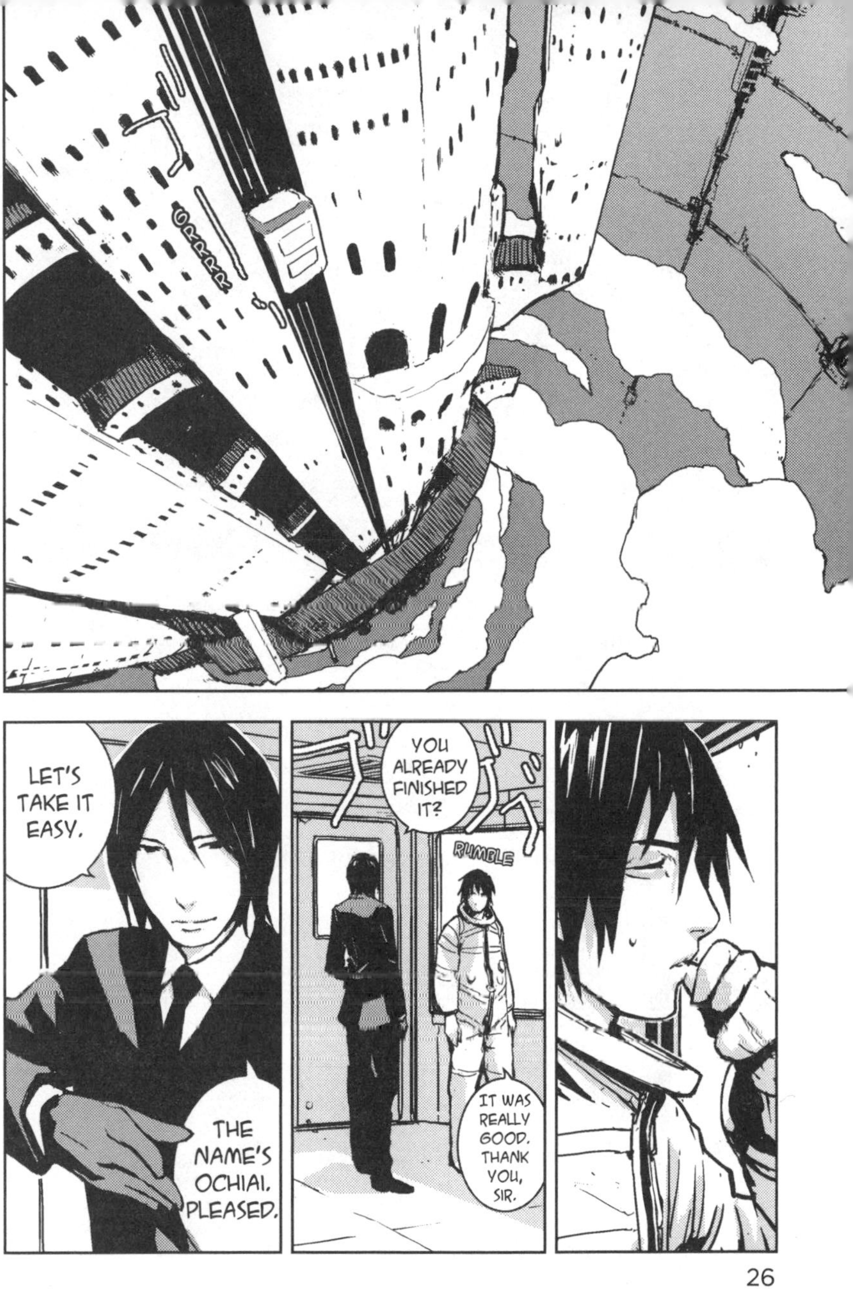
GRRRR
YOU ALREADY FINISHED IT?
RUMBLE
IT WAS REALLY GOOD. THANK YOU, SIR.
LET'S TAKE IT EASY.
THE NAME'S OCHIAI. PLEASED.

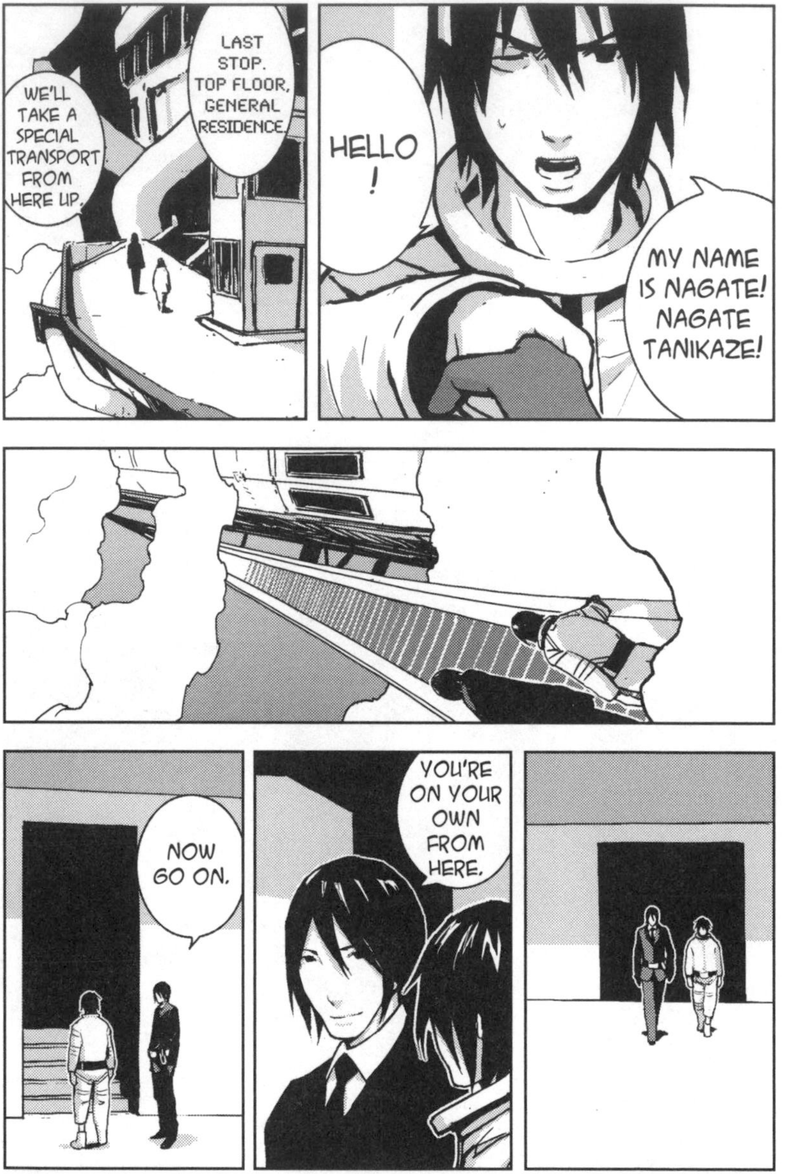
HELLO!
MY NAME IS NAGATE! NAGATE TANIKAZE!
LAST STOP. TOP FLOOR, GENERAL RESIDENCE.
WE'LL TAKE A SPECIAL TRANSPORT FROM HERE UP.
YOU'RE ON YOUR OWN FROM HERE.
NOW GO ON.

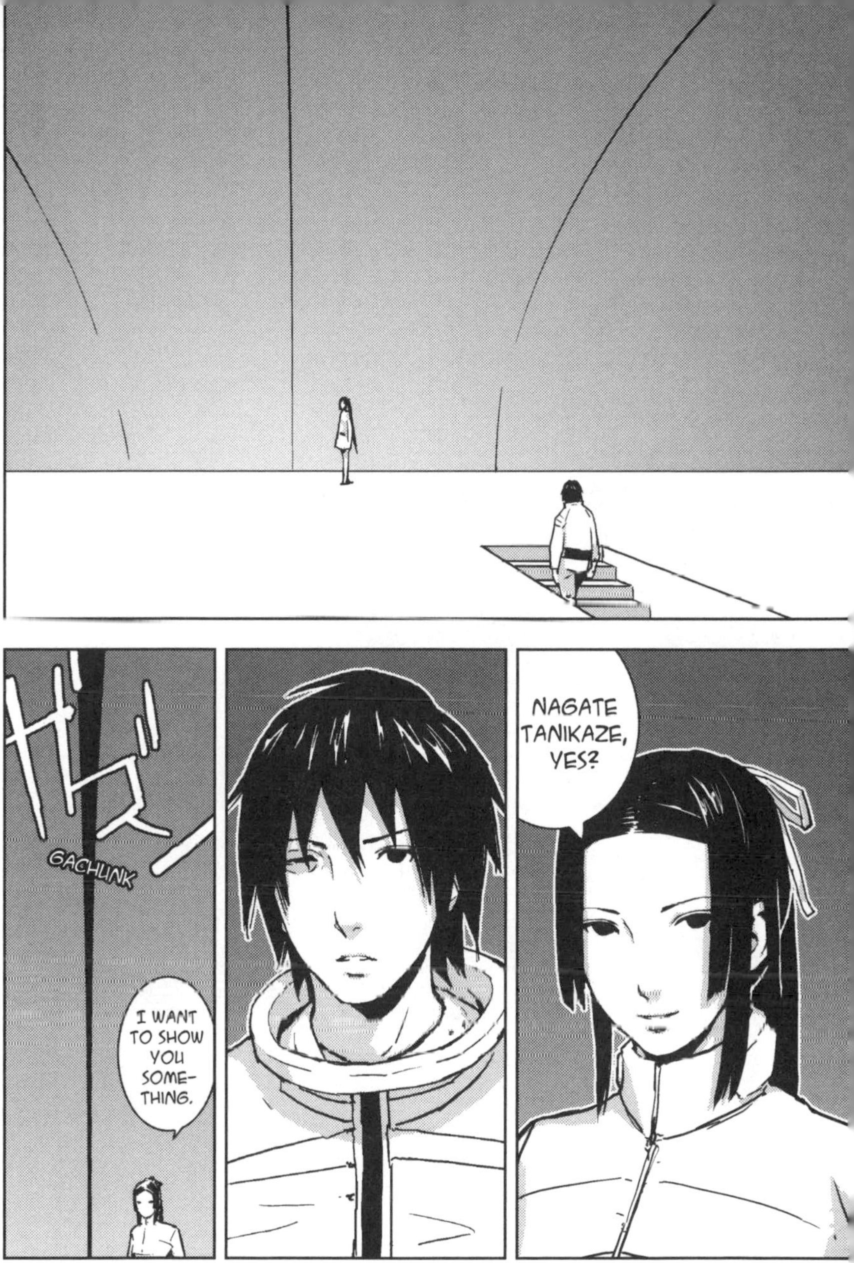
NAGATE TANIKAZE, YES?
GACHLINK
I WANT TO SHOW YOU SOME-THING.

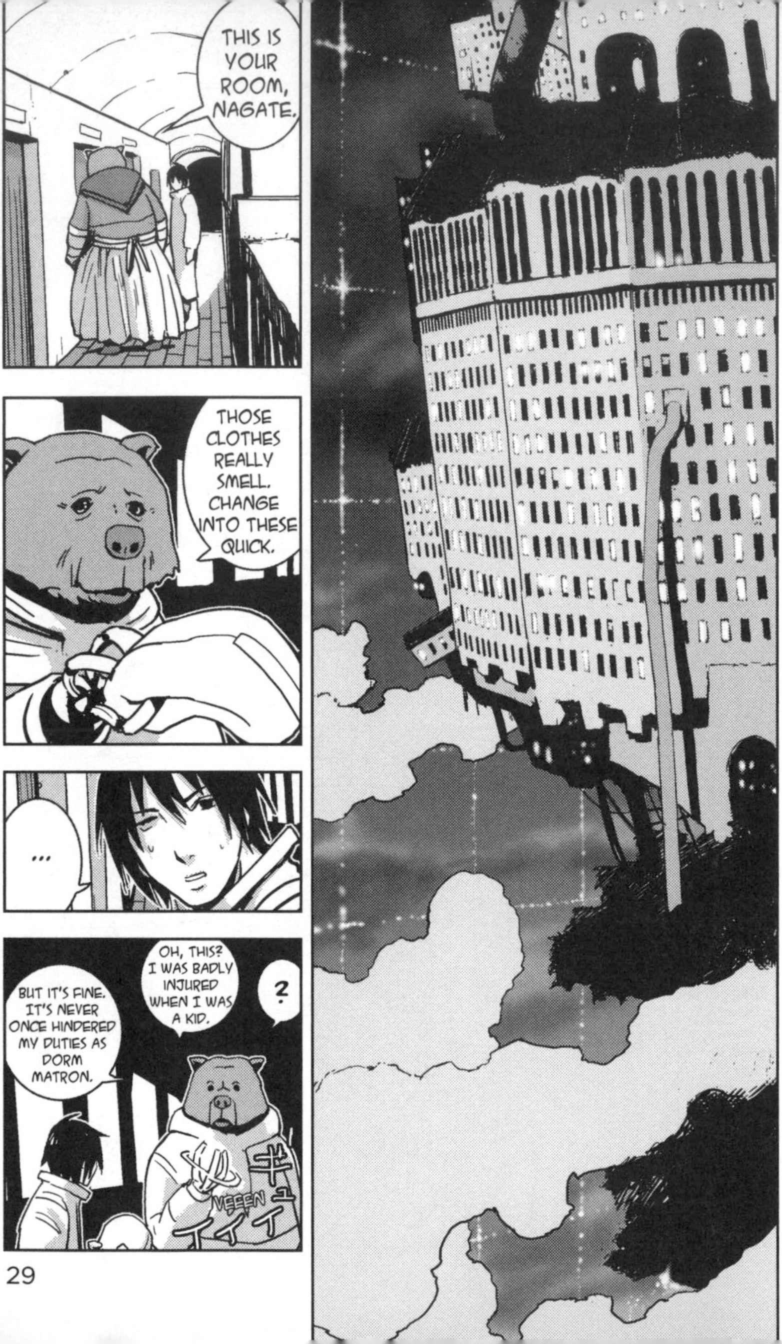
THIS IS YOUR ROOM, NAGATE.
THOSE CLOTHES REALLY SMELL. CHANGE INTO THESE QUICK.
...
?
OH, THIS? I WAS BADLY INJURED WHEN I WAS A KID.
BUT IT'S FINE. IT'S NEVER ONCE HINDERED MY DUTIES AS DORM MATRON.
VEEEN

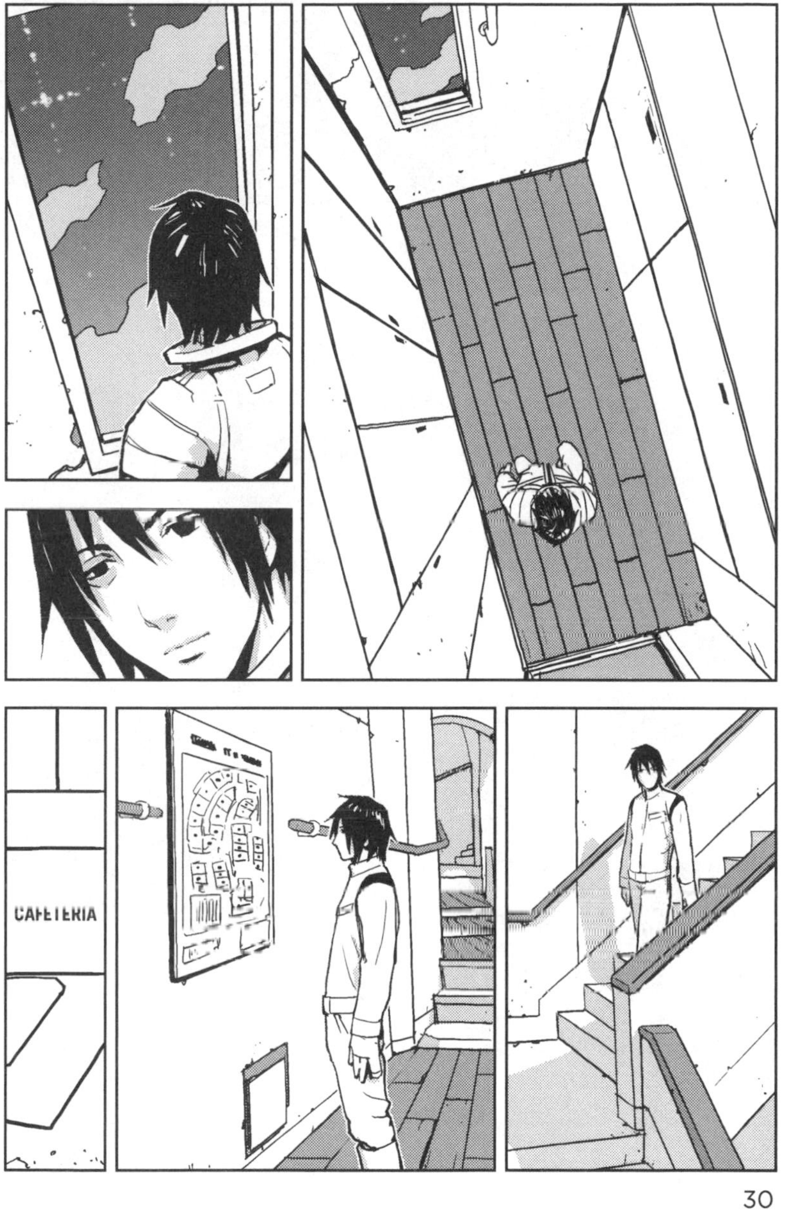
CAFETERIA

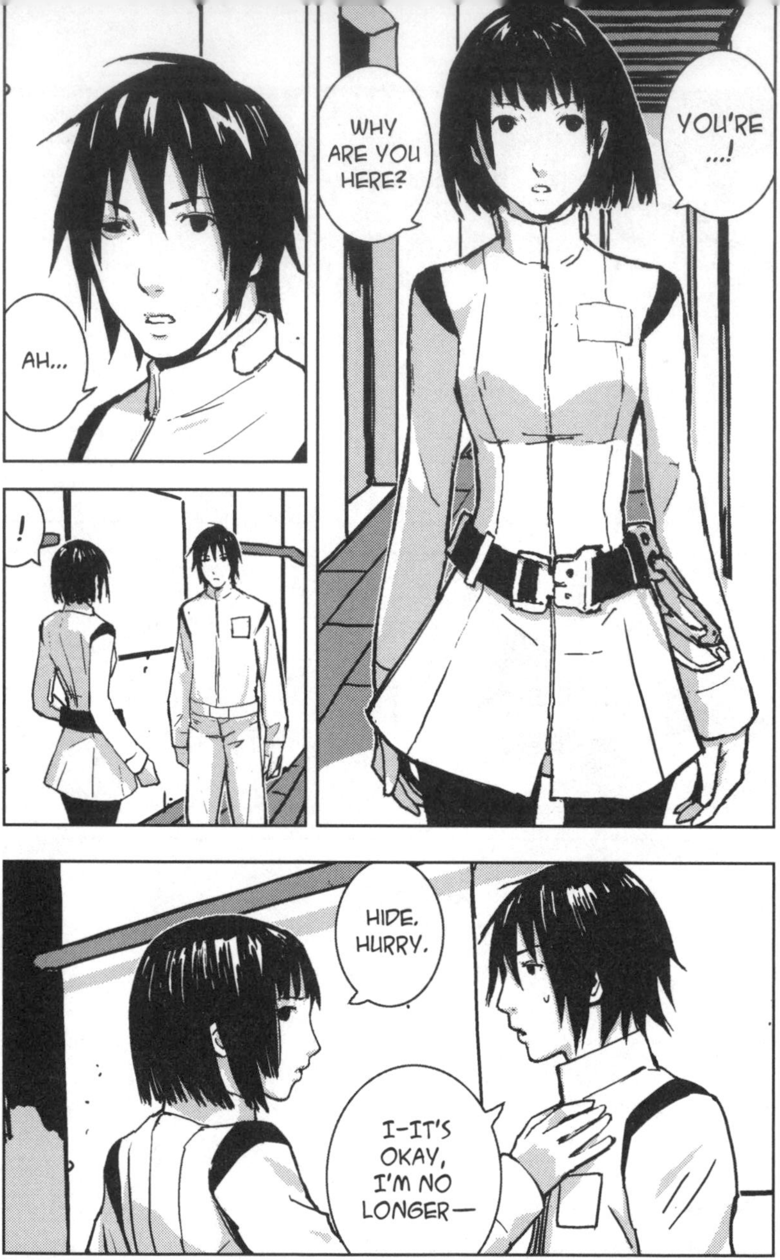
YOU'RE ...!
WHY ARE YOU HERE?
AH...
!
HIDE, HURRY.
I-IT'S OKAY, I'M NO LONGER—

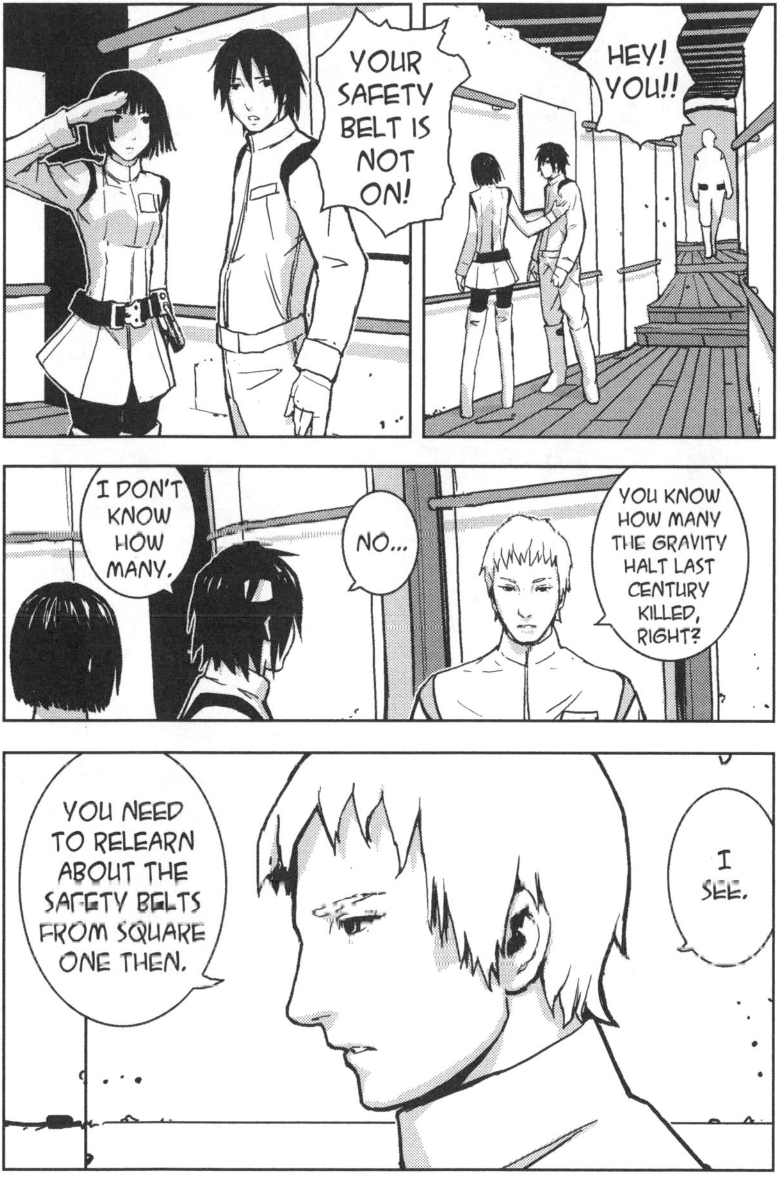
HEY! YOU!!
YOUR SAFETY BELT IS NOT ON!
YOU KNOW HOW MANY THE GRAVITY HALT LAST CENTURY KILLED, RIGHT?
NO...
I DON'T KNOW HOW MANY.
I SEE.
YOU NEED TO RELEARN ABOUT THE SAFETY BELTS FROM SQUARE ONE THEN.

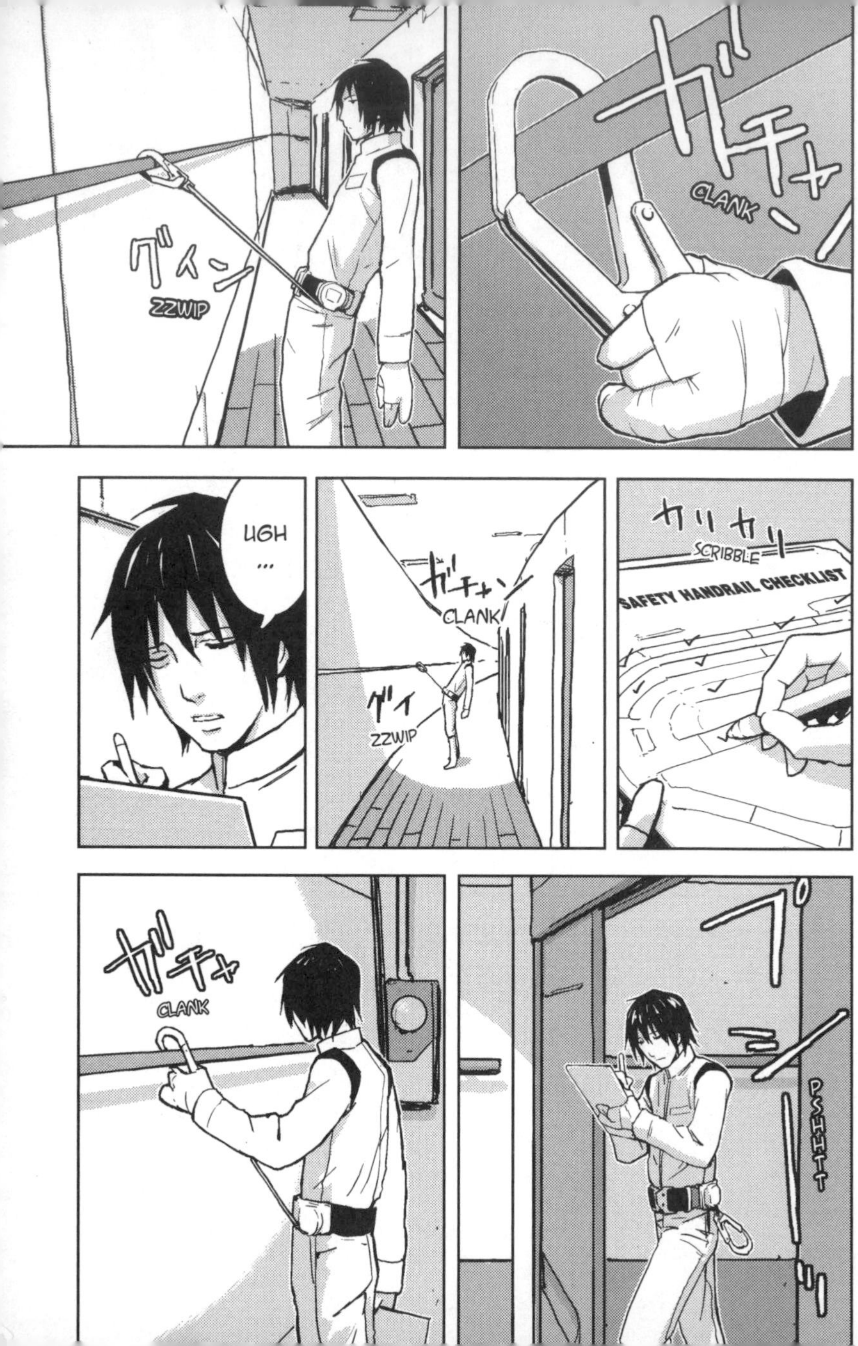
ガチャ
CLANK
グイン
ZZWIP
カリカリ
SCRIBBLE
SAFETY HANDRAIL CHECKLIST
ガチャ
CLANK
グイ
ZZWIP
UGH ...
PSHHTT
ガチャ
CLANK

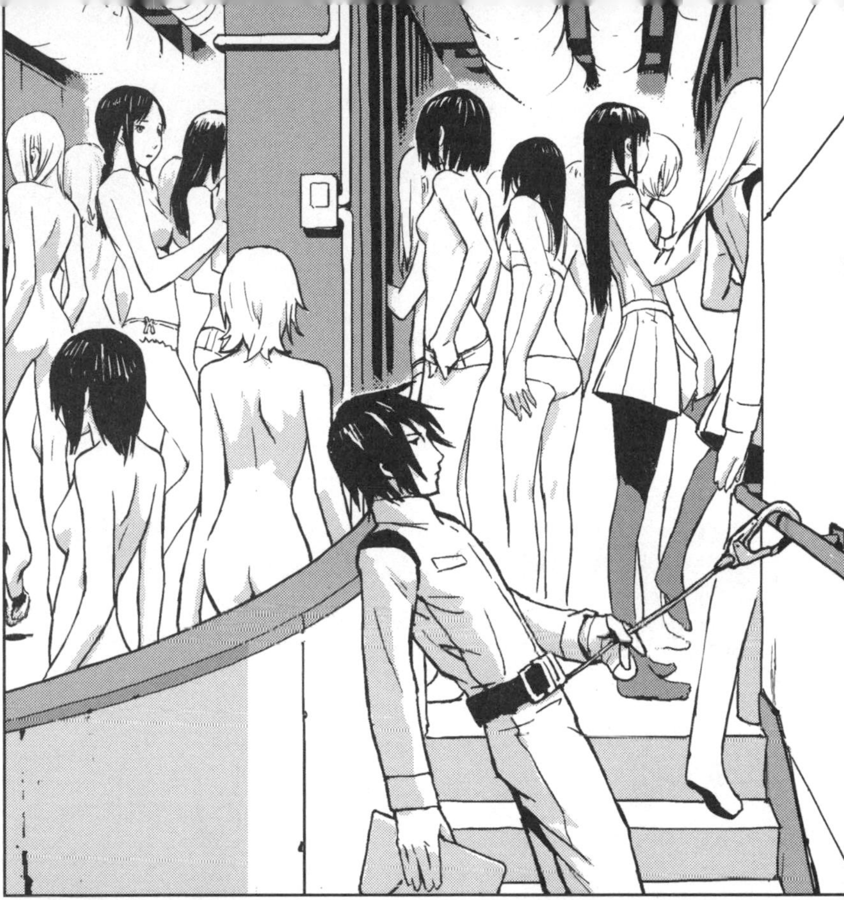

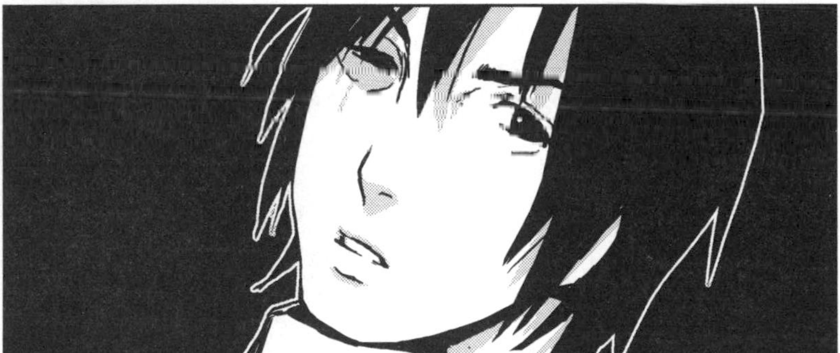

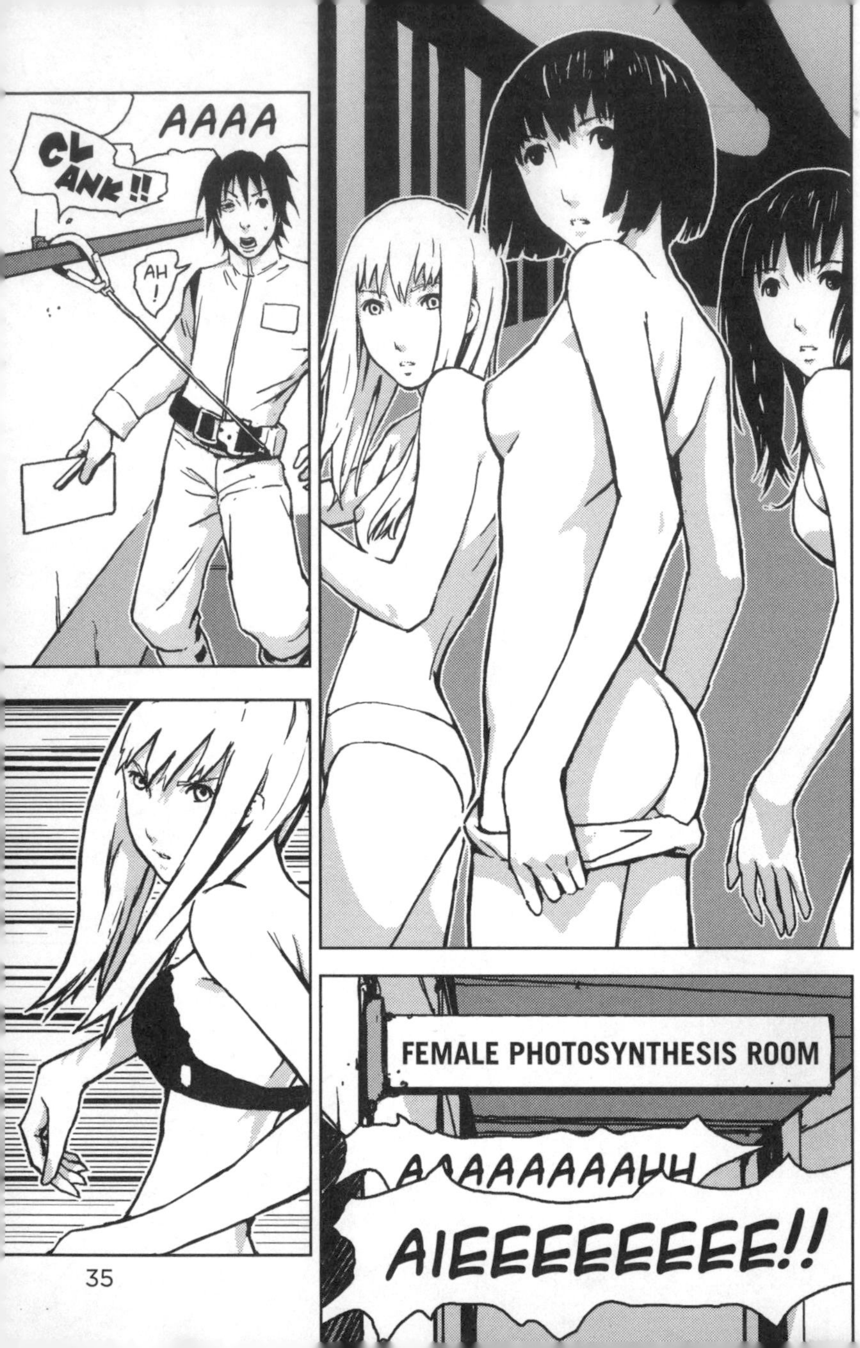
CLANK!!
AAAA
AH!
FEMALE PHOTOSYNTHESIS ROOM
AAAAAAAAHH
AIEEEEEEEEE!!

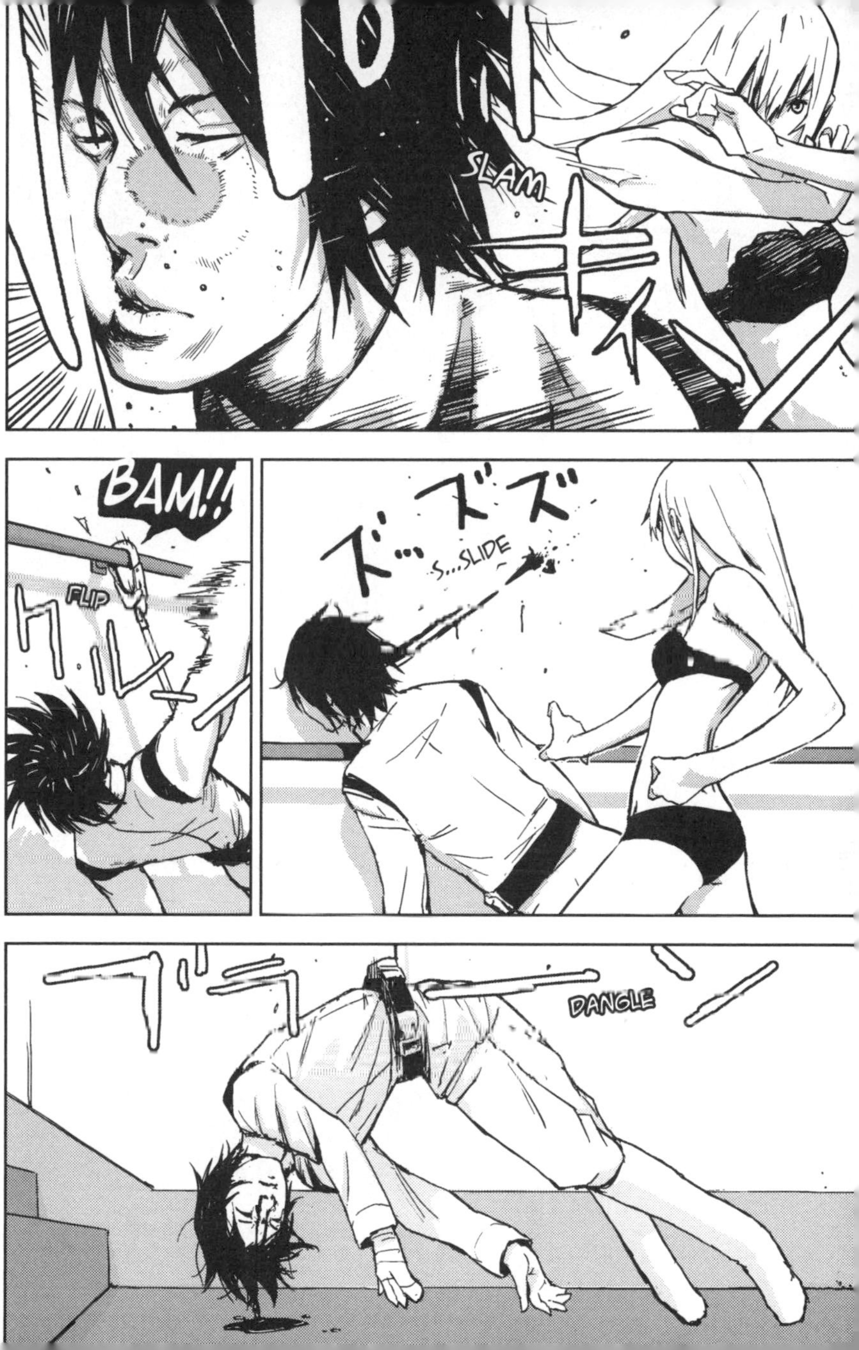
SLAM
ズ…ズズ
S...SLIDE
BAM!!
FLIP
DANGLE

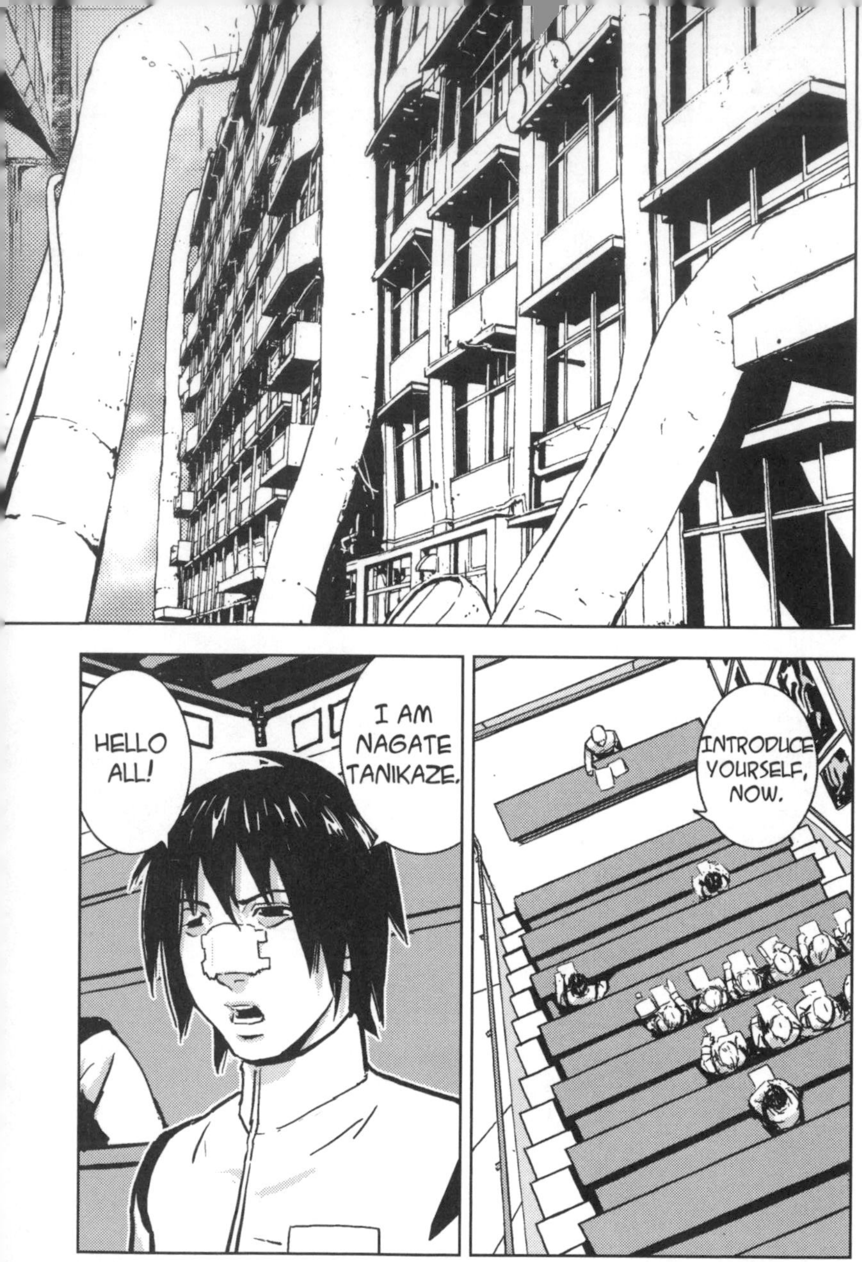
INTRODUCE YOURSELF, NOW.
I AM NAGATE TANIKAZE.
HELLO ALL!

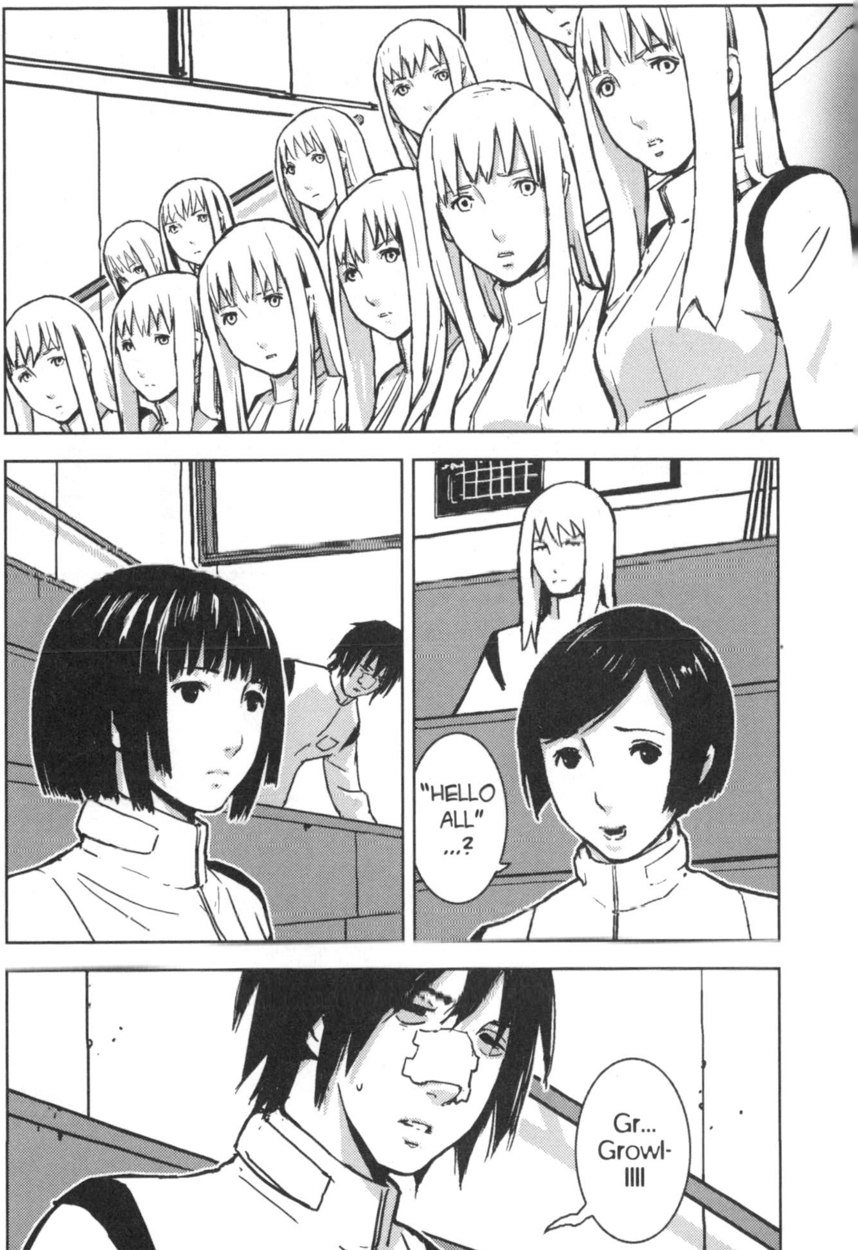
"HELLO ALL" ...?
Gr... Growl-llll

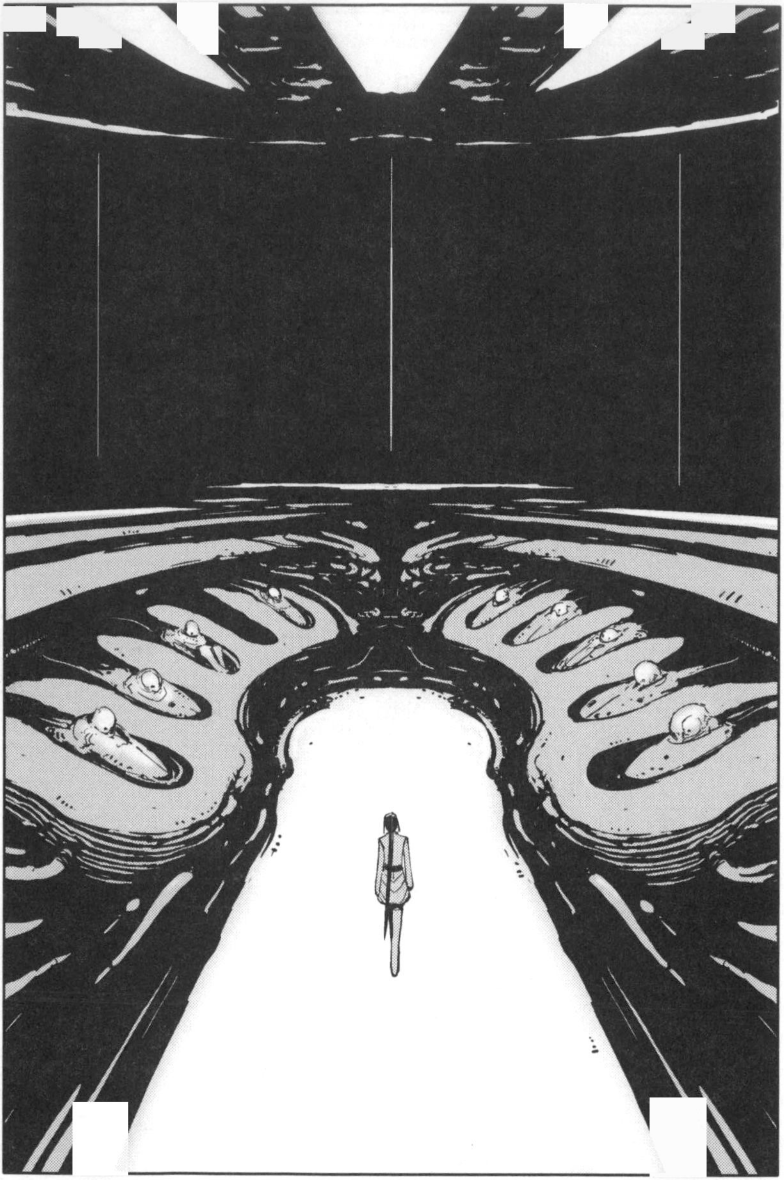

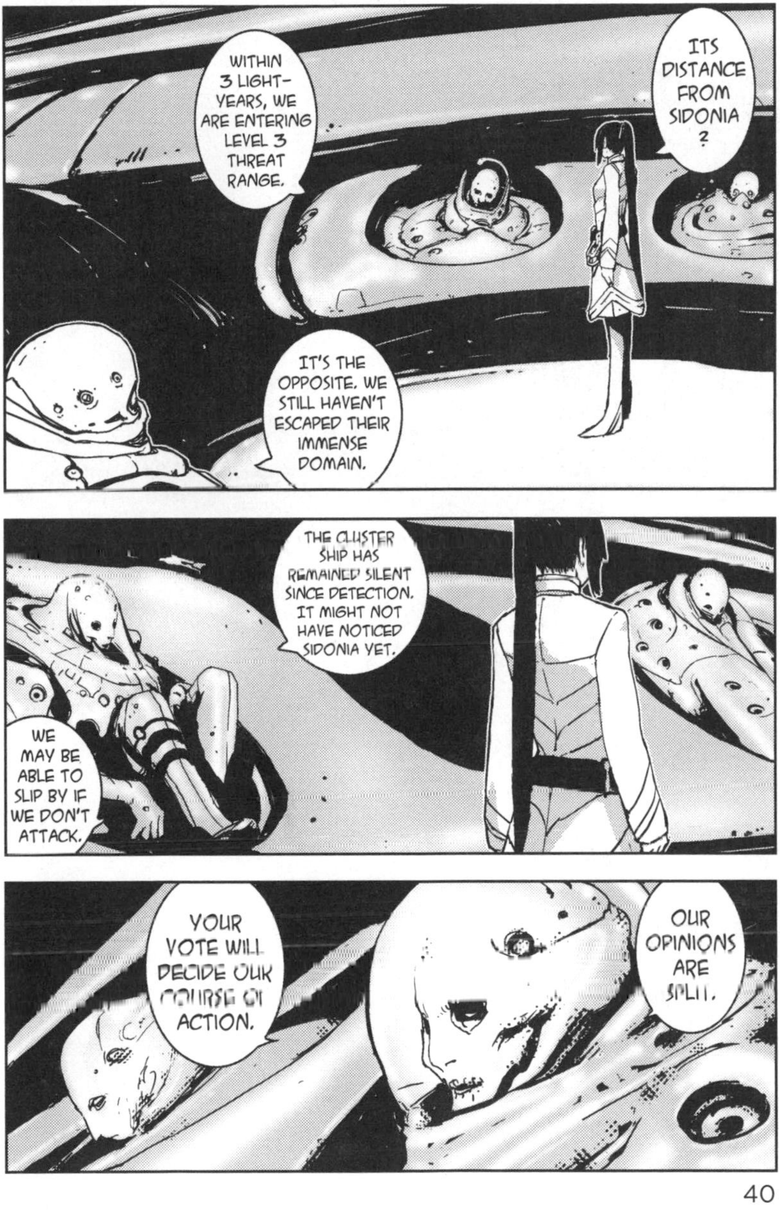
ITS DISTANCE FROM SIDONIA?
WITHIN 3 LIGHT-YEARS, WE ARE ENTERING LEVEL 3 THREAT RANGE.
IT'S THE OPPOSITE. WE STILL HAVEN'T ESCAPED THEIR IMMENSE DOMAIN.
THE CLUSTER SHIP HAS REMAINED SILENT SINCE DETECTION. IT MIGHT NOT HAVE NOTICED SIDONIA YET.
WE MAY BE ABLE TO SLIP BY IF WE DON'T ATTACK.
OUR OPINIONS ARE SPLIT.
YOUR VOTE WILL DECIDE OUR COURSE OF ACTION.

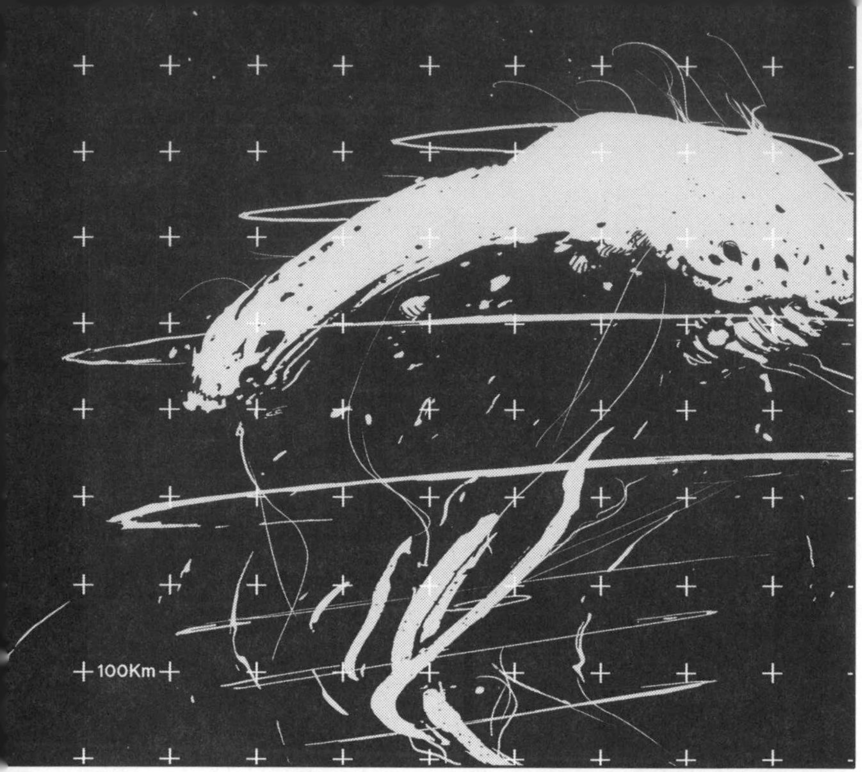
100Km

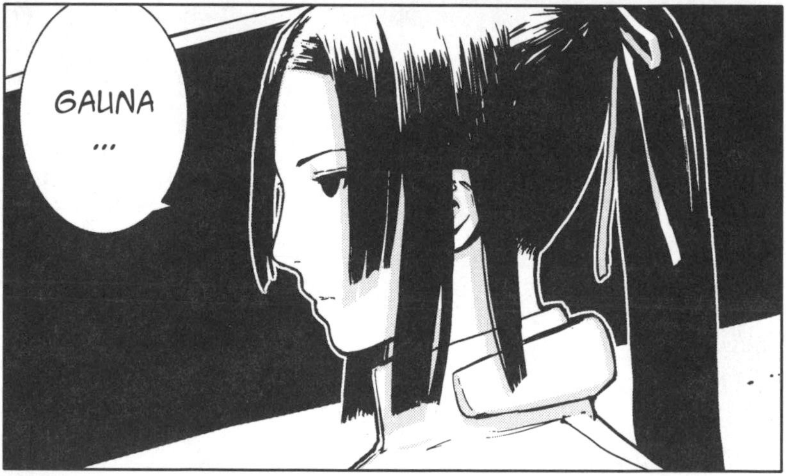
GAUNA
...

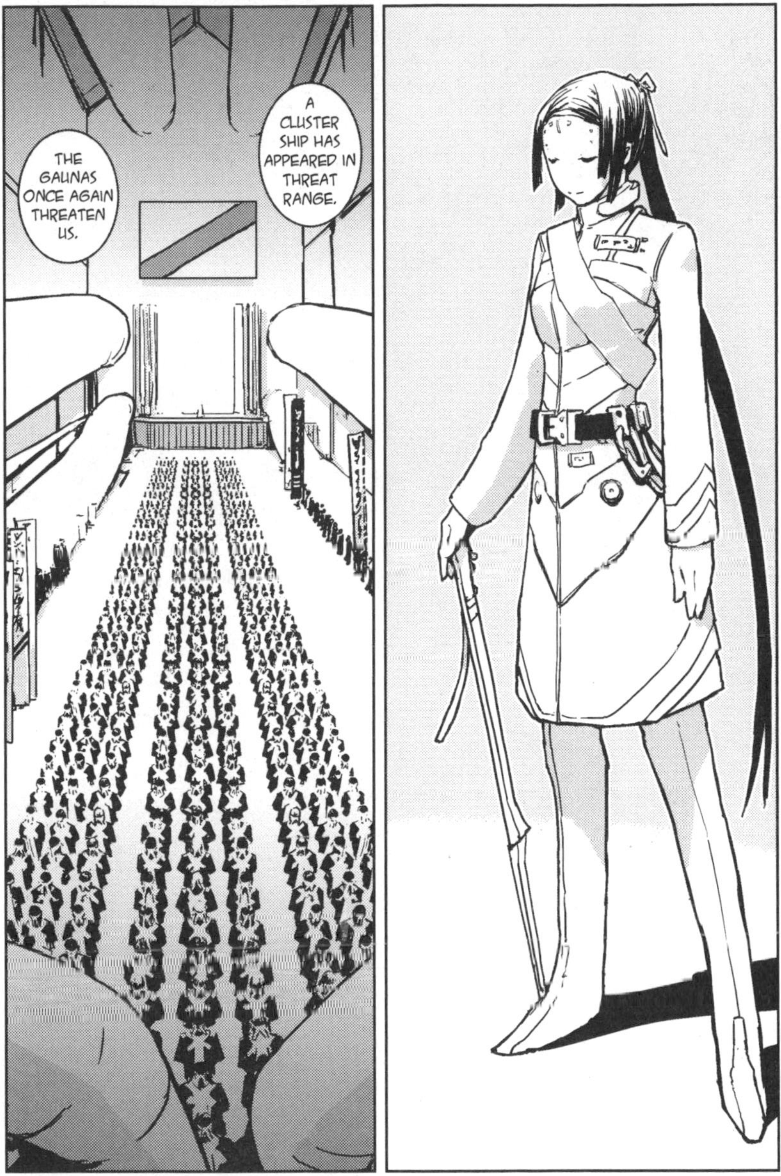
A CLUSTER SHIP HAS APPEARED IN THREAT RANGE.
THE GAUNAS ONCE AGAIN THREATEN US.

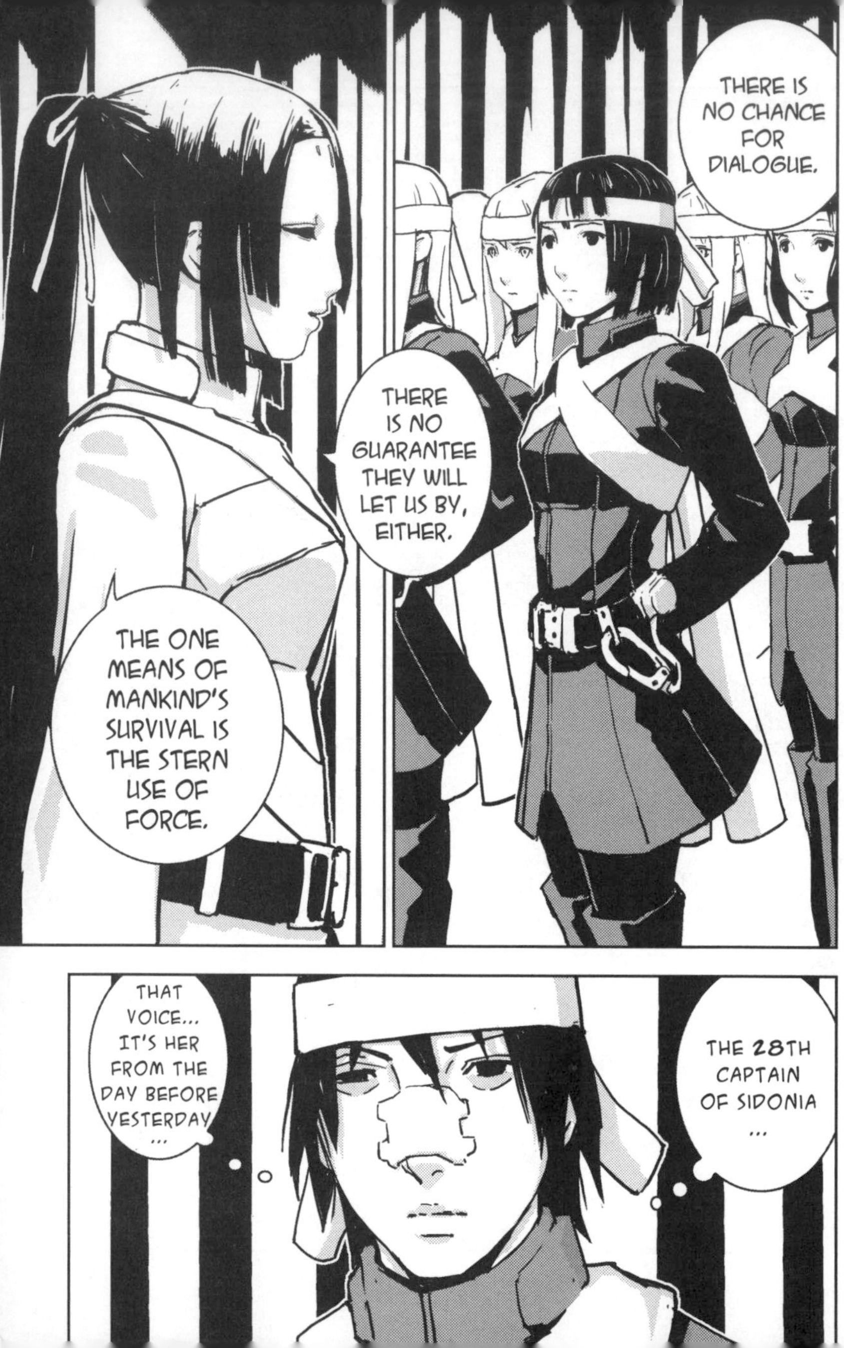

THERE IS NO CHANCE FOR DIALOGUE.
THERE IS NO GUARANTEE THEY WILL LET US BY, EITHER.
THE ONE MEANS OF MANKIND'S SURVIVAL IS THE STERN USE OF FORCE.
THE 28TH CAPTAIN OF SIDONIA ...
THAT VOICE... IT'S HER FROM THE DAY BEFORE YESTERDAY ...

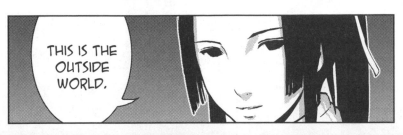

THIS IS THE
OUTSIDE
WORLD.

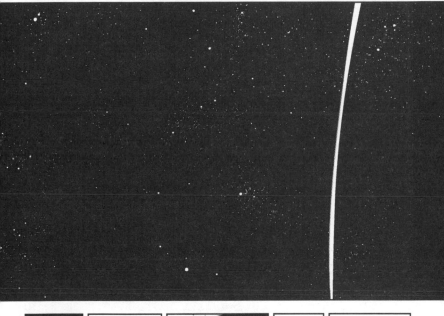

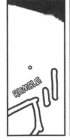

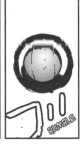

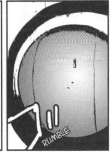

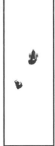

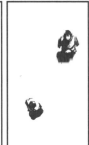

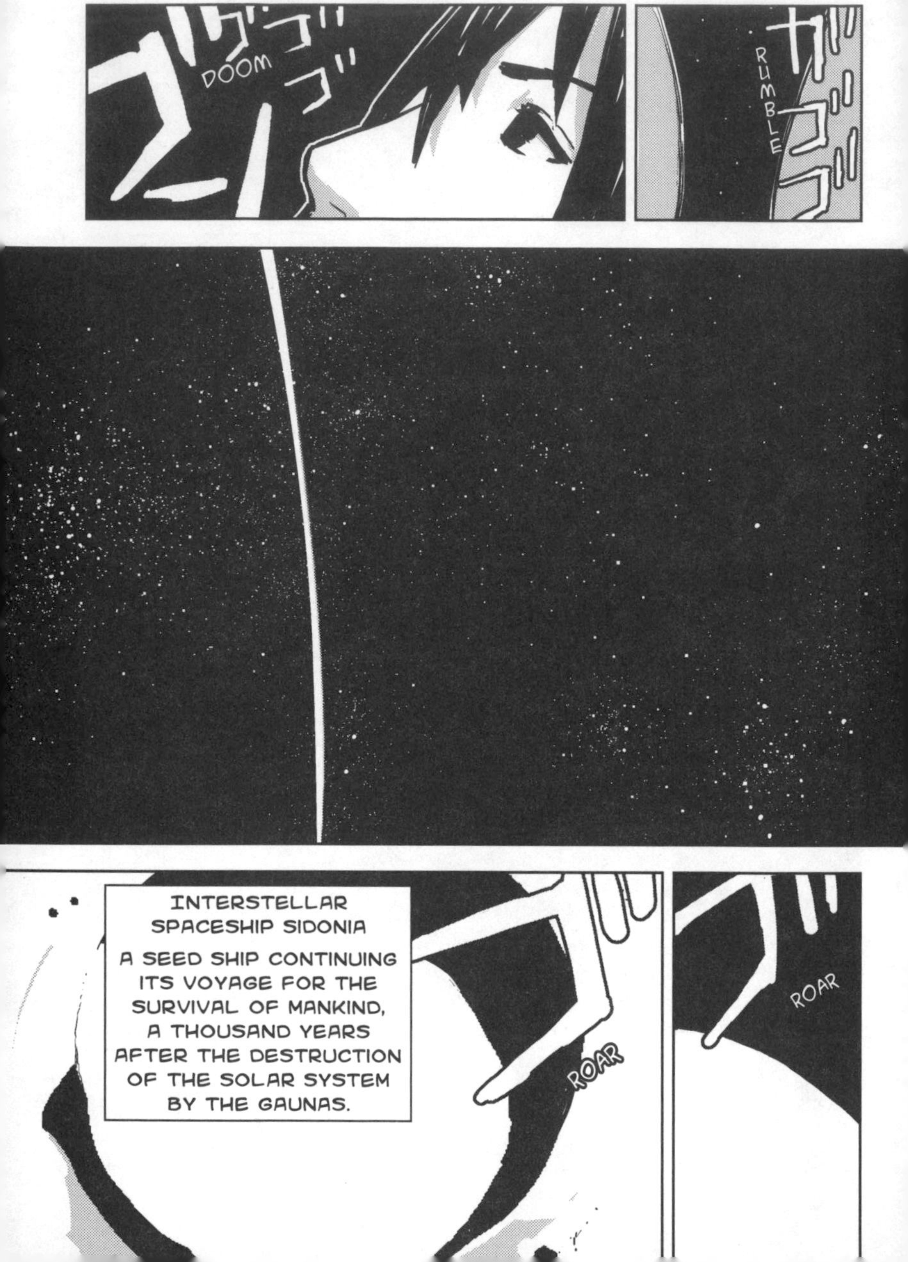
RUMBLE
DOOM
ROAR
INTERSTELLAR SPACESHIP SIDONIA
A SEED SHIP CONTINUING ITS VOYAGE FOR THE SURVIVAL OF MANKIND, A THOUSAND YEARS AFTER THE DESTRUCTION OF THE SOLAR SYSTEM BY THE GAUNAS.
ROAR

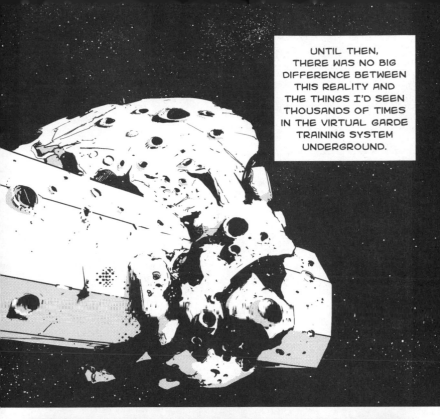

UNTIL THEN,
THERE WAS NO BIG
DIFFERENCE BETWEEN
THIS REALITY AND
THE THINGS I'D SEEN
THOUSANDS OF TIMES
IN THE VIRTUAL GARDE
TRAINING SYSTEM
UNDERGROUND.

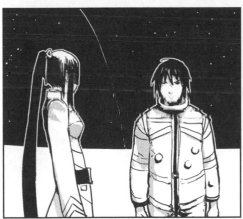

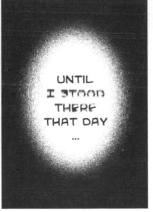

UNTIL
I STOOD
THERE
THAT DAY
...

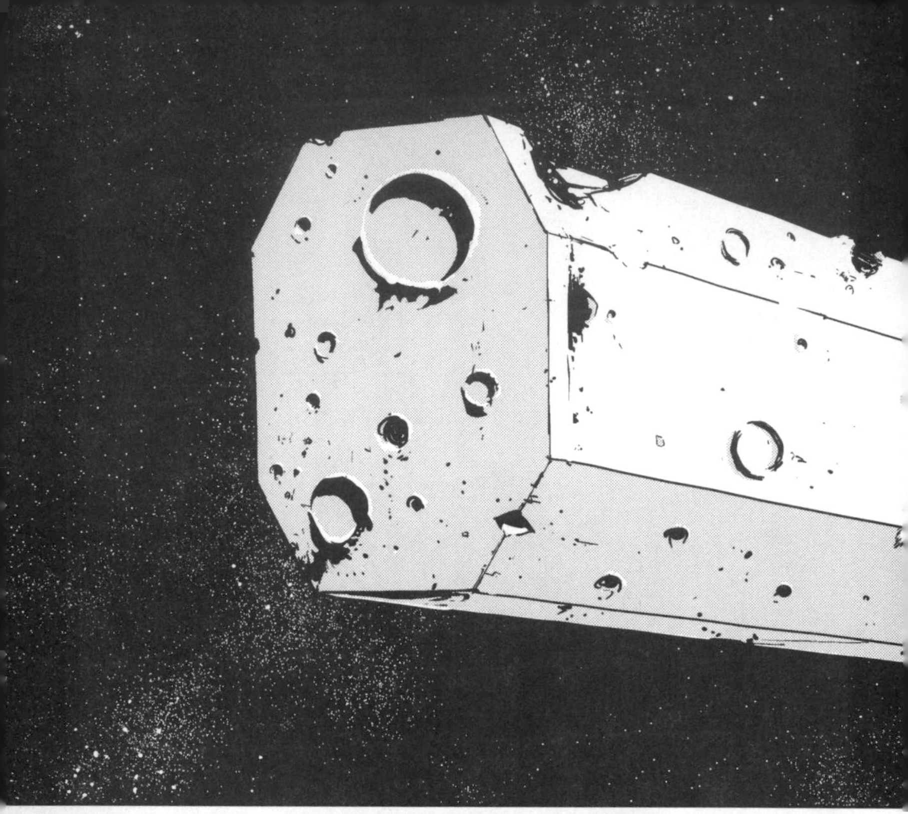

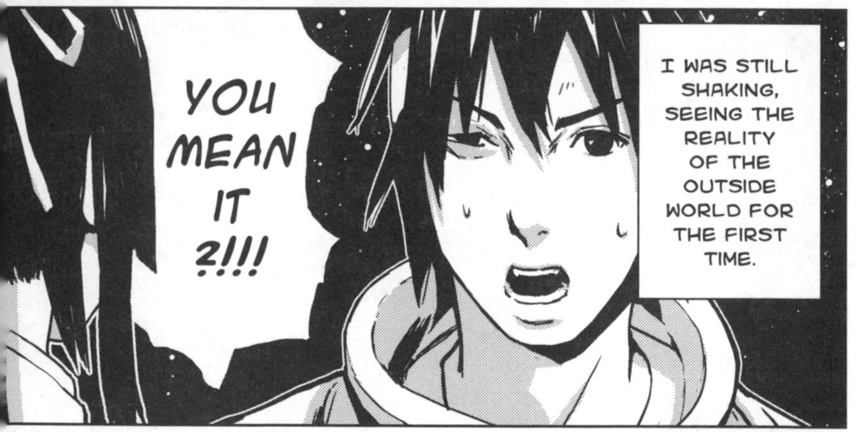
I WAS STILL SHAKING, SEEING THE REALITY OF THE OUTSIDE WORLD FOR THE FIRST TIME.
YOU MEAN IT ?!!!

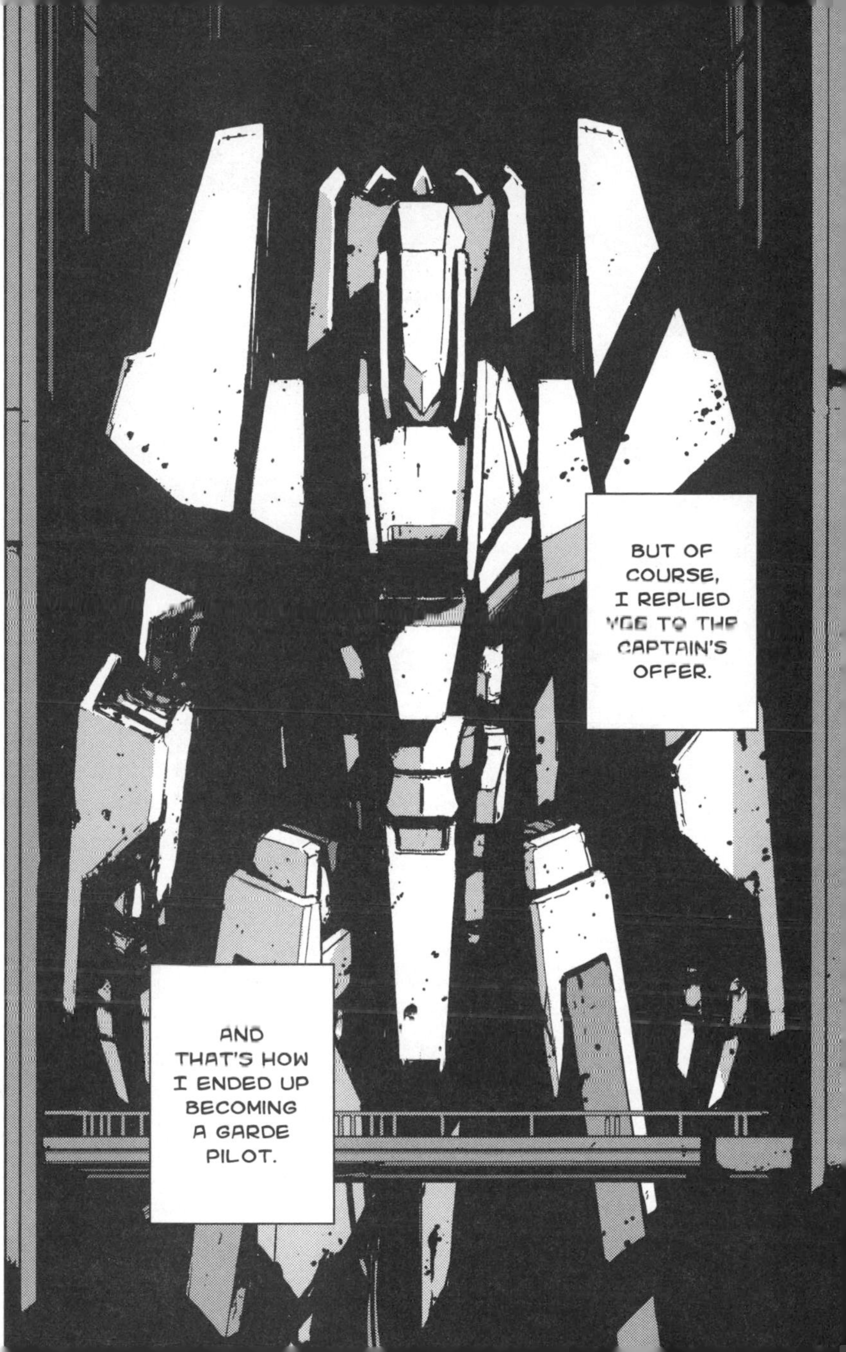
BUT OF COURSE, I REPLIED YES TO THE CAPTAIN'S OFFER.
AND THAT'S HOW I ENDED UP BECOMING A GARDE PILOT.

Chapter 1: END

Chapter 2: Nagate Tanikaze's Maiden Battle

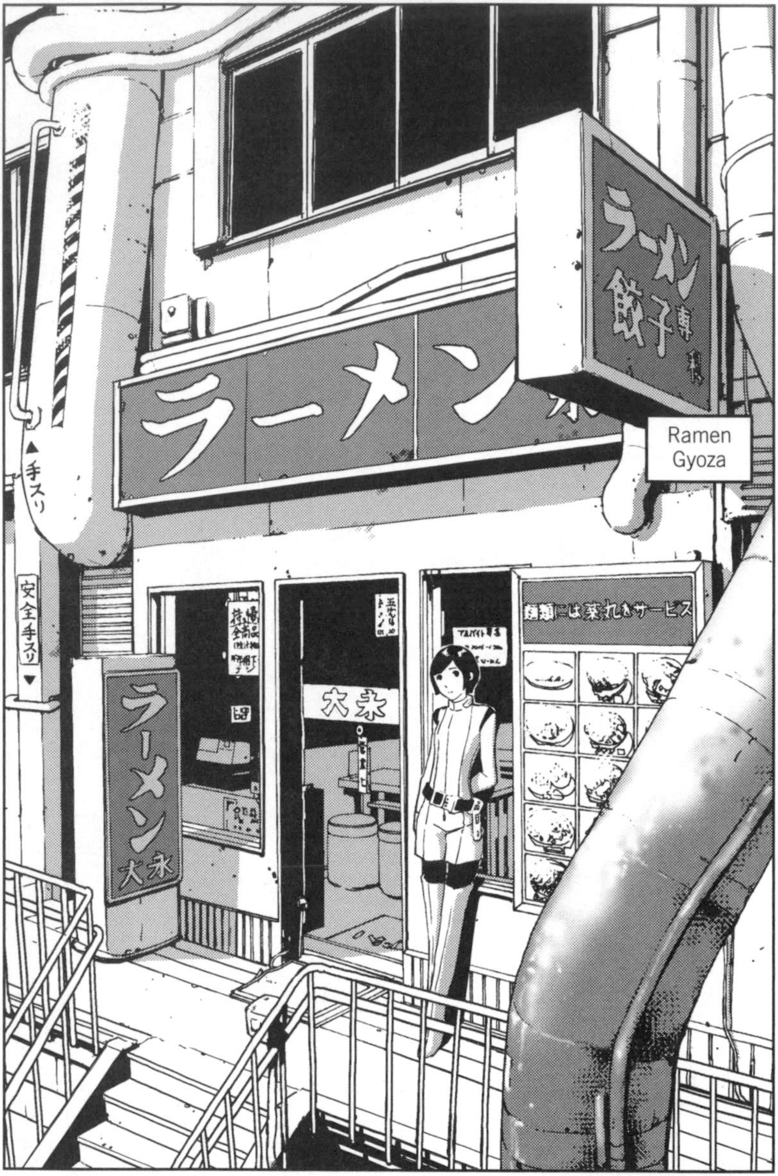
ラーメン
ラーメン
餃子專
Ramen
Gyoza
▲手スリ
安全手スリ▼
ラーメン
大永
大永

One Hundred Sights of Sidonia Part One: Residential District ①

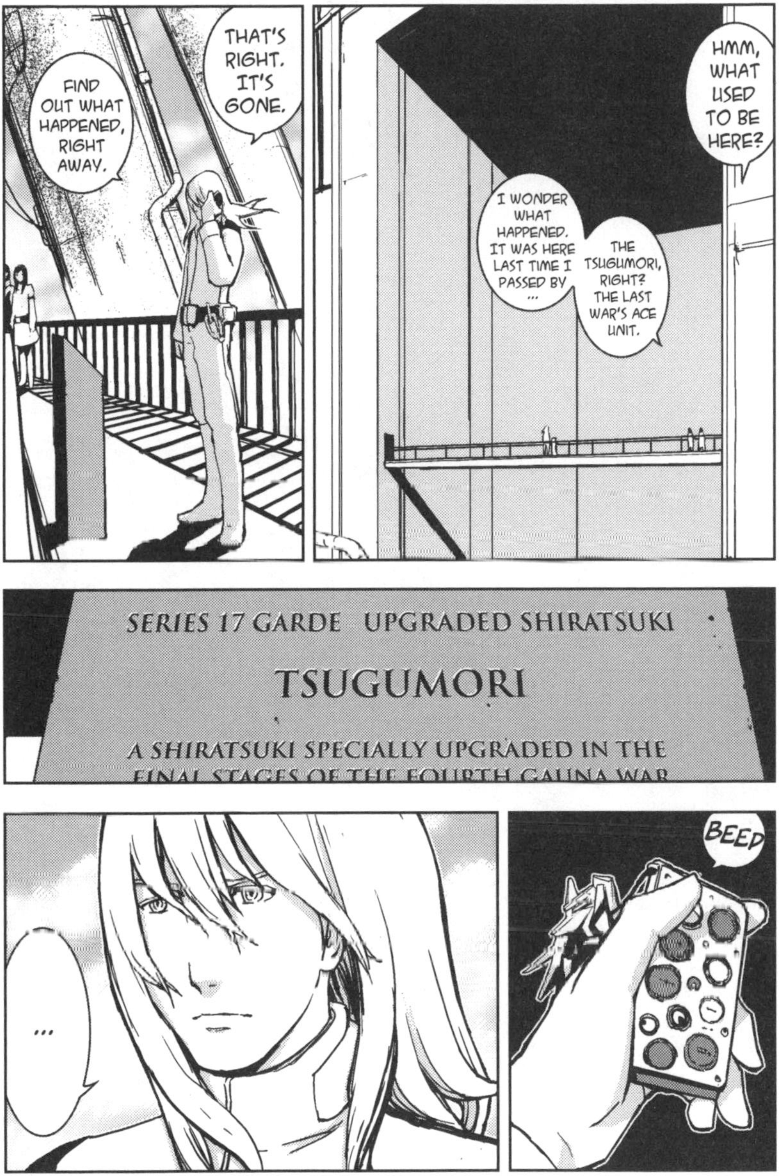
HMM, WHAT USED TO BE HERE?
THE TSUGUMORI, RIGHT? THE LAST WAR'S ACE UNIT.
I WONDER WHAT HAPPENED. IT WAS HERE LAST TIME I PASSED BY ...
THAT'S RIGHT. IT'S GONE.
FIND OUT WHAT HAPPENED, RIGHT AWAY.
SERIES 17 GARDE UPGRADED SHIRATSUKI
TSUGUMORI
A SHIRATSUKI SPECIALLY UPGRADED IN THE
BEED
...

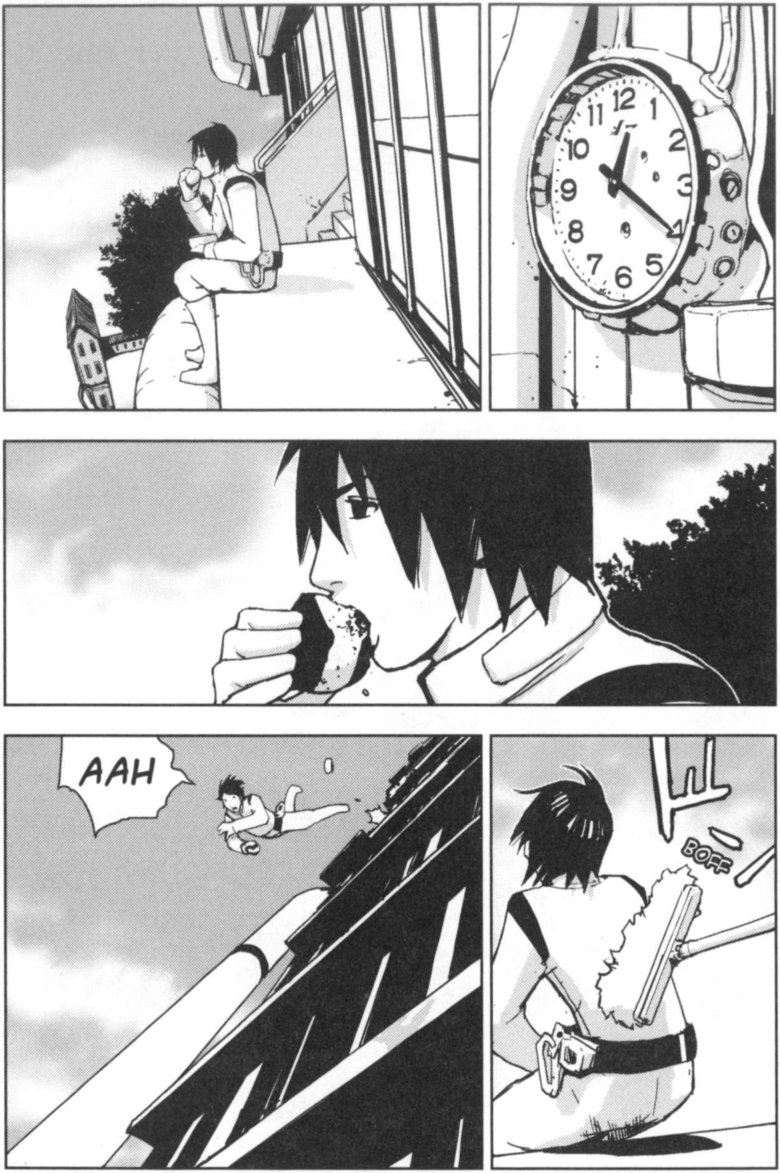
BOFF
AAH

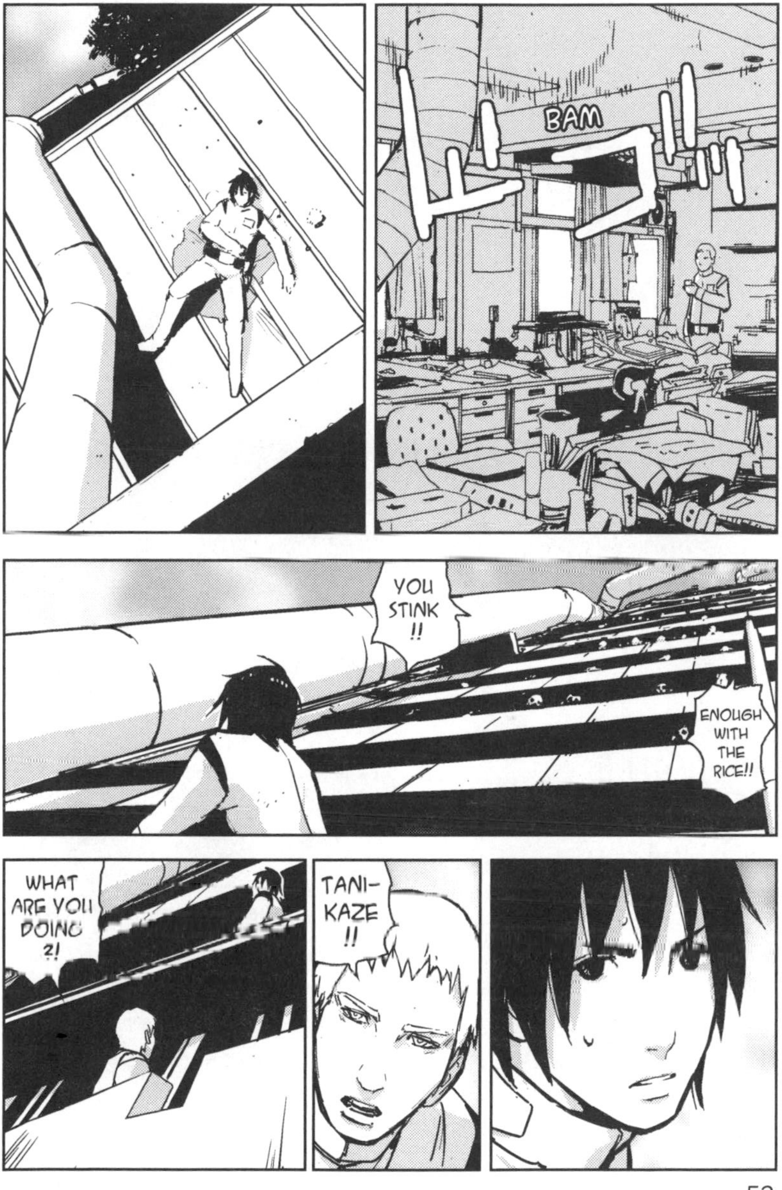
BAM
YOU STINK!!
ENOUGH WITH THE RICE!!
TANI-KAZE!!
WHAT ARE YOU DOING?!

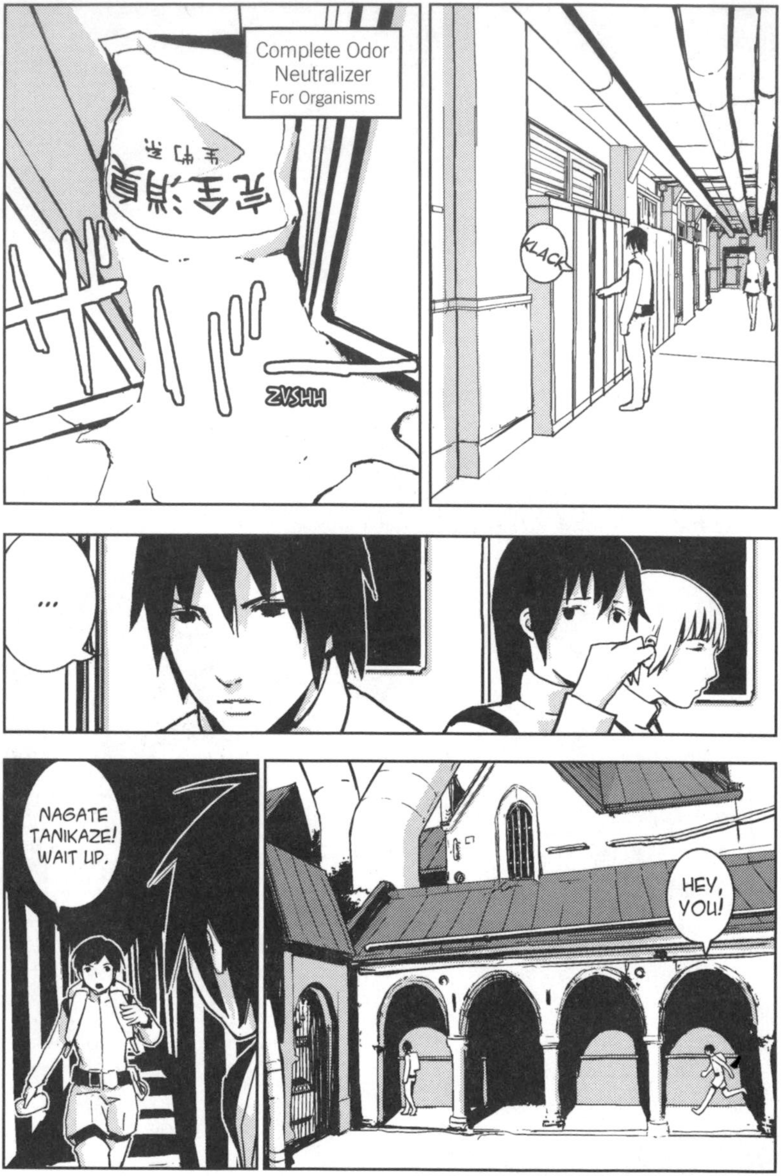
KLACK
Complete Odor Neutralizer For Organisms
ZVSHH
...
HEY, YOU!
NAGATE TANIKAZE! WAIT UP.

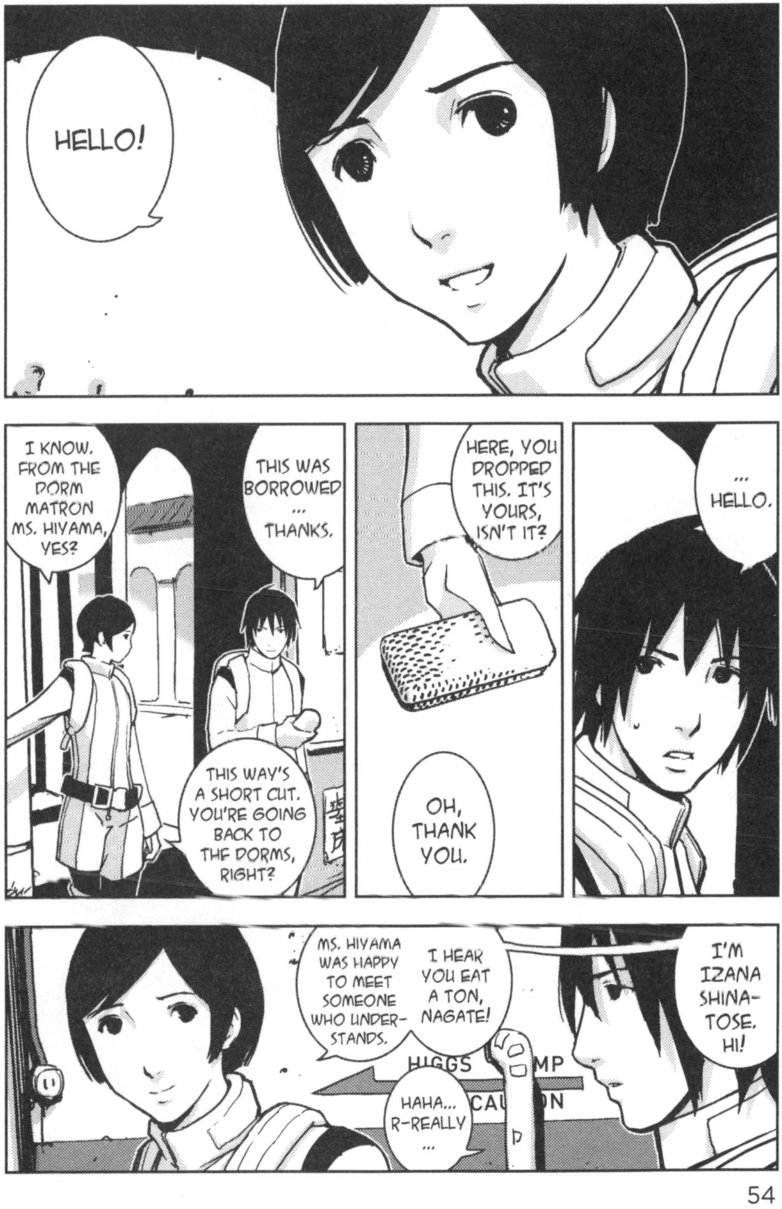
HELLO!
... HELLO.
HERE, YOU DROPPED THIS. IT'S YOURS, ISN'T IT?
OH, THANK YOU.
THIS WAS BORROWED ... THANKS.
I KNOW. FROM THE DORM MATRON MS. HIYAMA, YES?
THIS WAY'S A SHORT CUT. YOU'RE GOING BACK TO THE DORMS, RIGHT?
I'M IZANA SHINATOSE. HI!
I HEAR YOU EAT A TON, NAGATE!
MS. HIYAMA WAS HAPPY TO MEET SOMEONE WHO UNDERSTANDS.
HAHA... R-REALLY ...

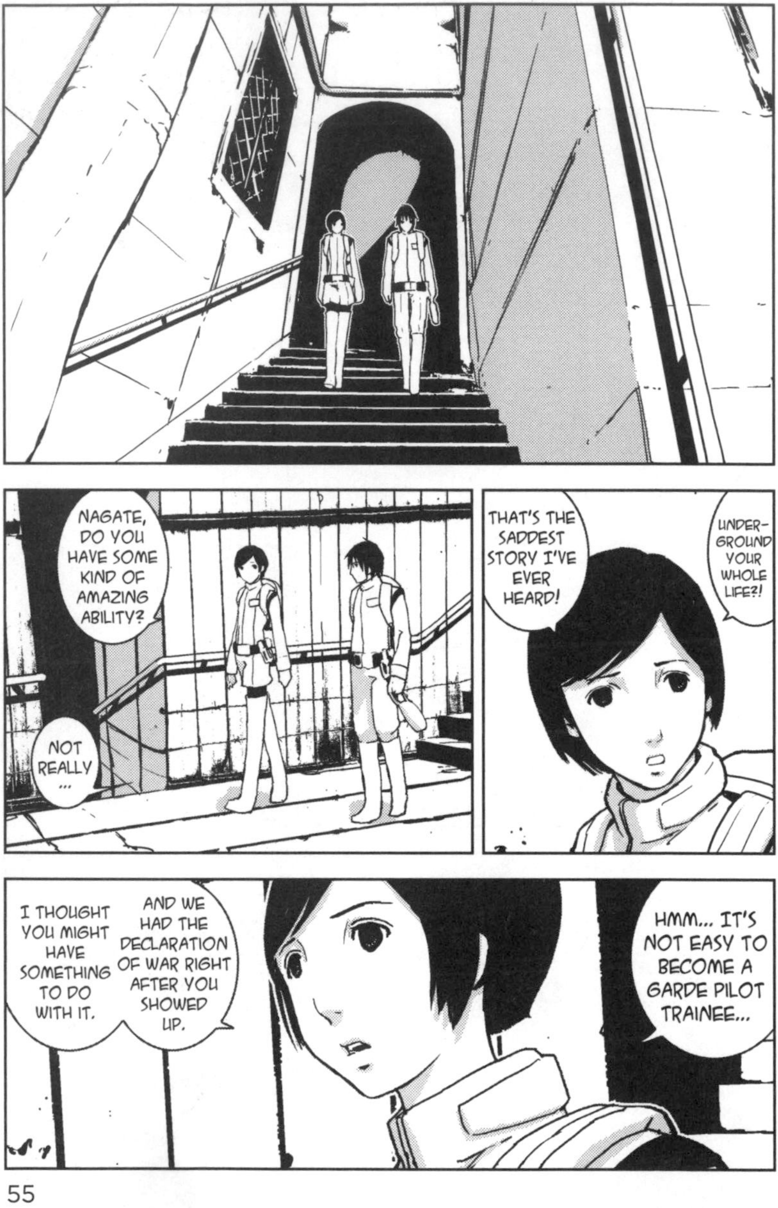
UNDER-GROUND YOUR WHOLE LIFE?!
THAT'S THE SADDEST STORY I'VE EVER HEARD!
NAGATE, DO YOU HAVE SOME KIND OF AMAZING ABILITY?
NOT REALLY ...
HMM... IT'S NOT EASY TO BECOME A GARDE PILOT TRAINEE...
AND WE HAD THE DECLARATION OF WAR RIGHT AFTER YOU SHOWED UP.
I THOUGHT YOU MIGHT HAVE SOMETHING TO DO WITH IT.

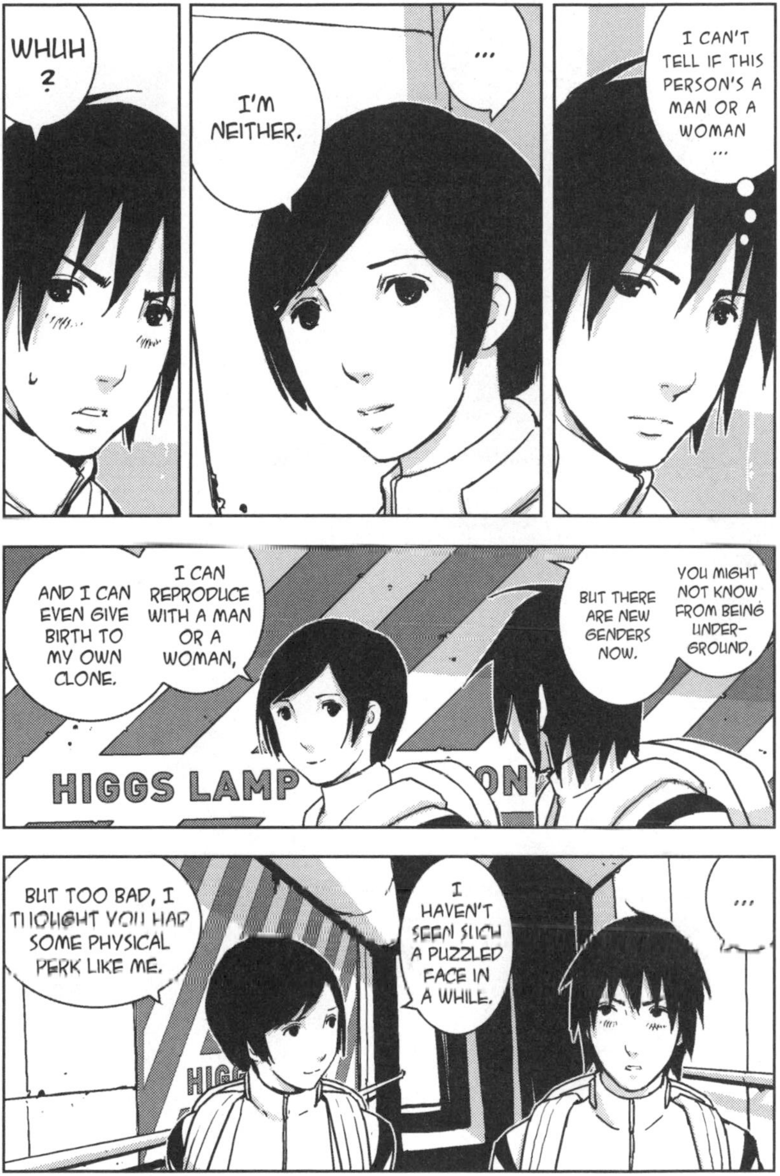
I CAN'T TELL IF THIS PERSON'S A MAN OR A WOMAN ...
...
I'M NEITHER.
WHUH ?
YOU MIGHT NOT KNOW FROM BEING UNDER-GROUND,
BUT THERE ARE NEW GENDERS NOW.
I CAN REPRODUCE WITH A MAN OR A WOMAN,
AND I CAN EVEN GIVE BIRTH TO MY OWN CLONE.
HIGGS LAMP
ON
...
I HAVEN'T SEEN SUCH A PUZZLED FACE IN A WHILE.
BUT TOO BAD, I THOUGHT YOU HAD SOME PHYSICAL PERK LIKE ME.

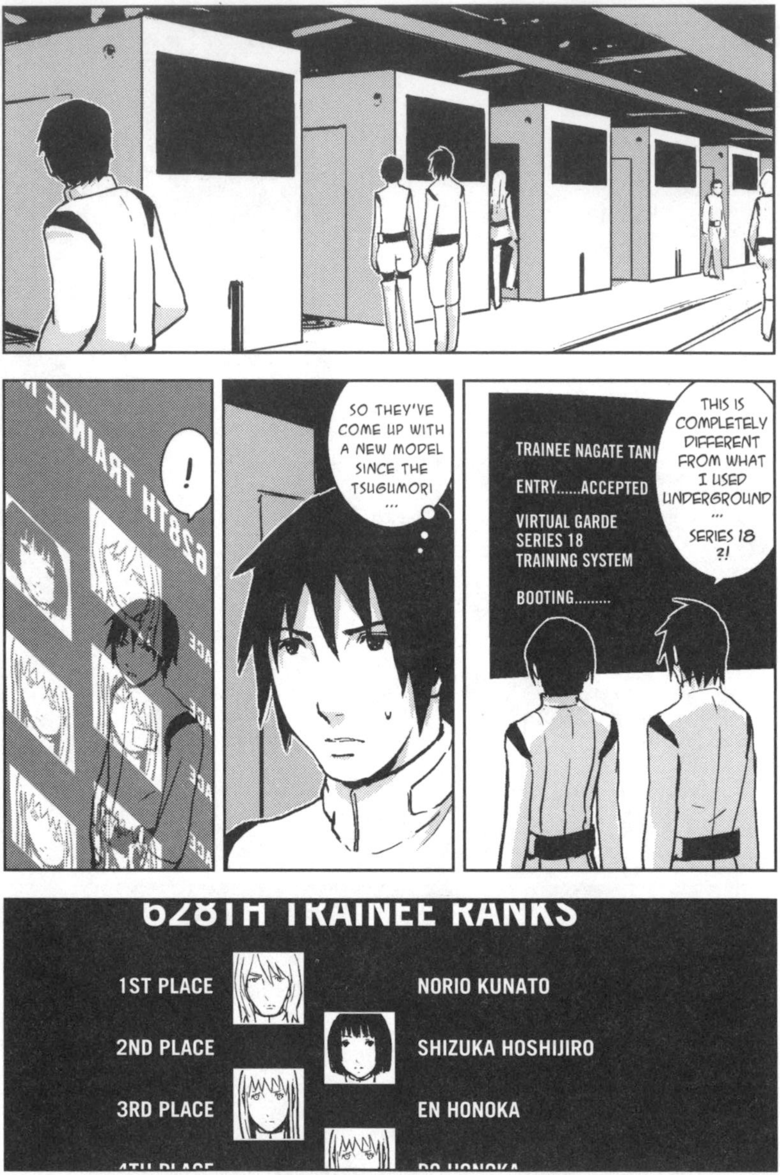
TRAINEE NAGATE TANI
ENTRY......ACCEPTED
VIRTUAL GARDE SERIES 18 TRAINING SYSTEM
BOOTING.........
THIS IS COMPLETELY DIFFERENT FROM WHAT I USED UNDERGROUND ... SERIES 18 ?!
SO THEY'VE COME UP WITH A NEW MODEL SINCE THE TSUGUMORI ...
!
628TH TRAINEE RANKS
1ST PLACE
NORIO KUNATO
2ND PLACE
SHIZUKA HOSHIJIRO
3RD PLACE
EN HONOKA

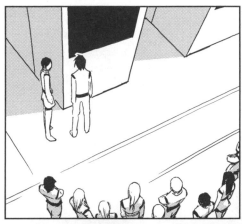
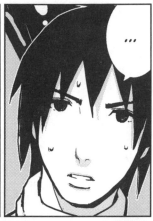
...

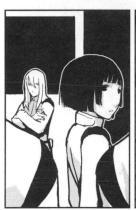

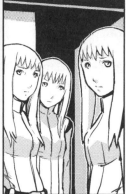

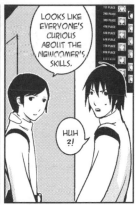

LOOKS LIKE EVERYONE'S CURIOUS ABOUT THE NEWCOMER'S SKILLS.

| 1ST PLACE |
| 2ND PLACE |
| 3RD PLACE |
| 4TH PLACE |
| 5TH PLACE |
| 6TH PLACE |
| 7TH PLACE |
| 8TH PLACE |

HUH ?!

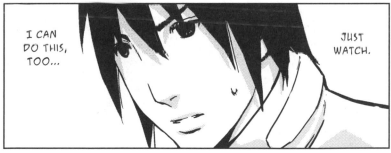

I CAN DO THIS, TOO...

JUST WATCH.

58

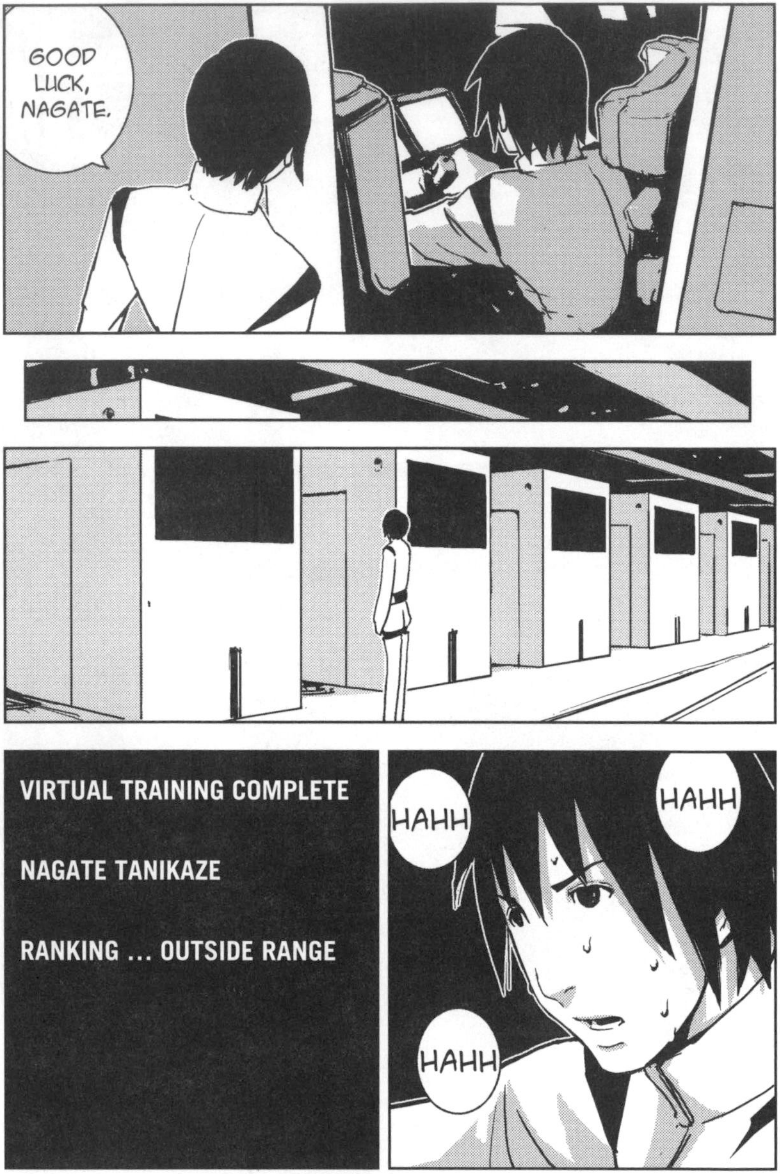
GOOD LUCK, NAGATE.
HAHH
HAHH
HAHH
VIRTUAL TRAINING COMPLETE
NAGATE TANIKAZE
RANKING ... OUTSIDE RANGE

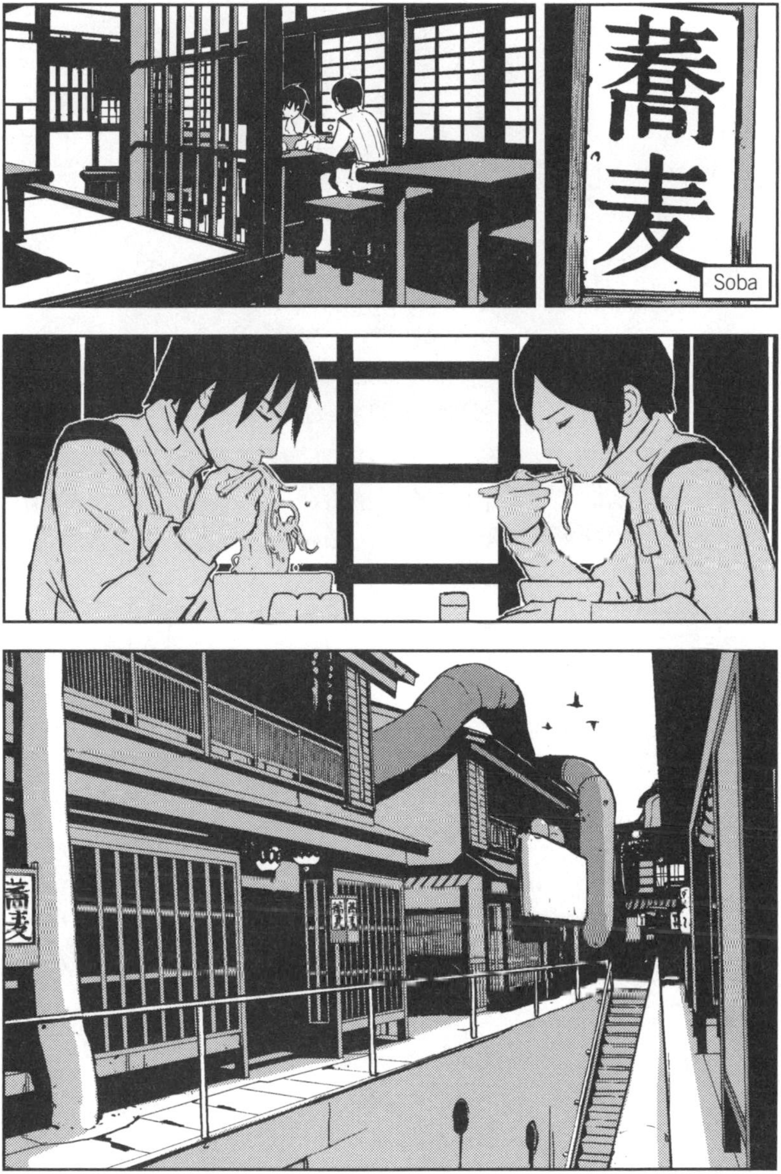
蕎麦
Soba
蕎麦

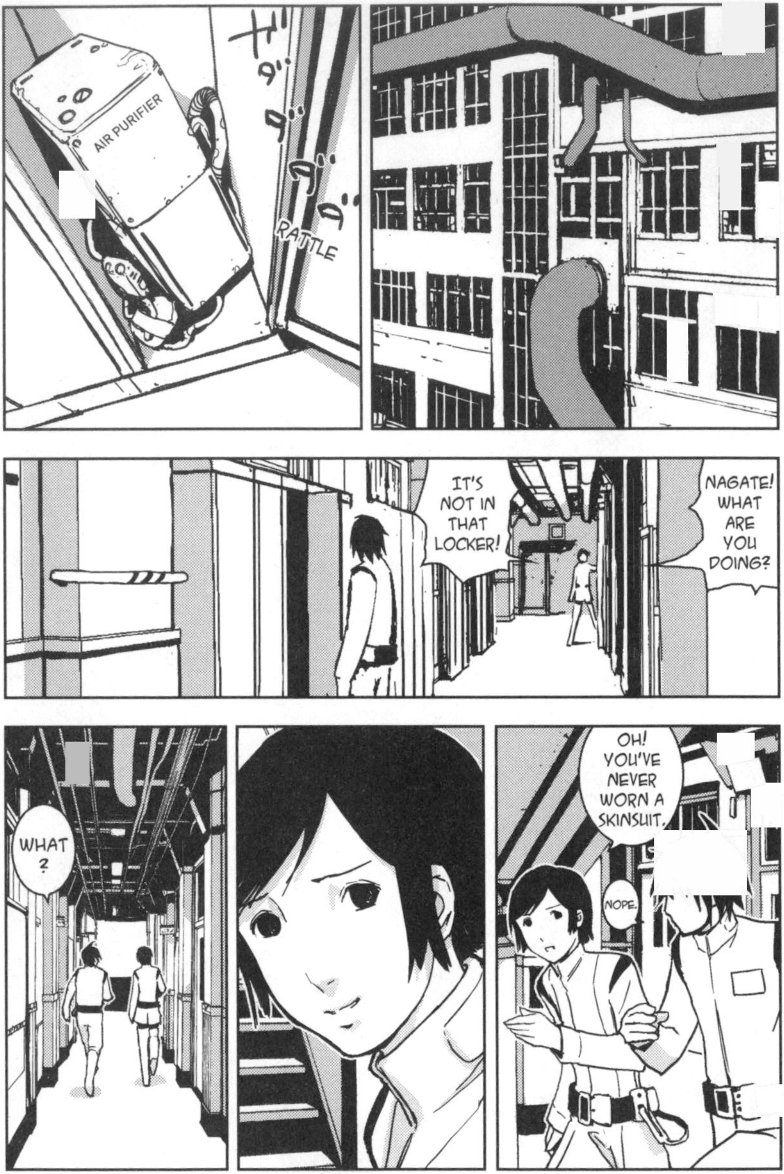
AIR PURIFIER
RATTLE
NAGATE! WHAT ARE YOU DOING?
IT'S NOT IN THAT LOCKER!
OH! YOU'VE NEVER WORN A SKINSUIT.
NOPE.
WHAT?

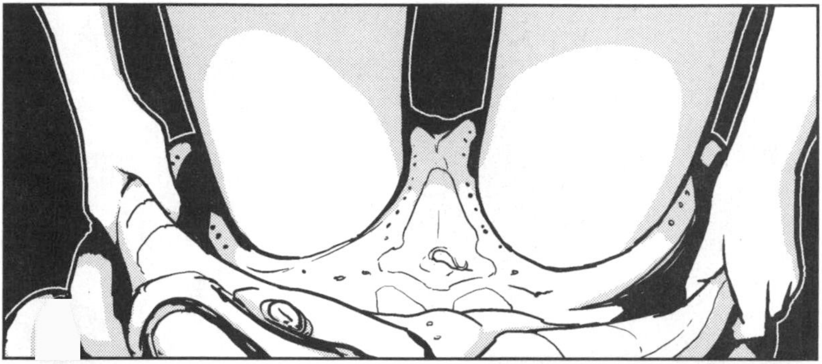

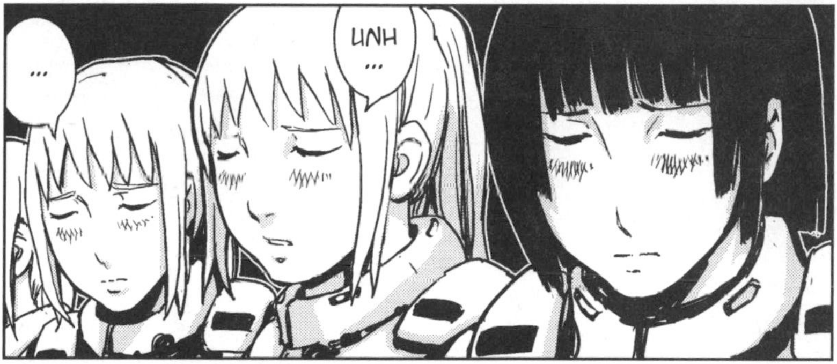
UNH
...
...

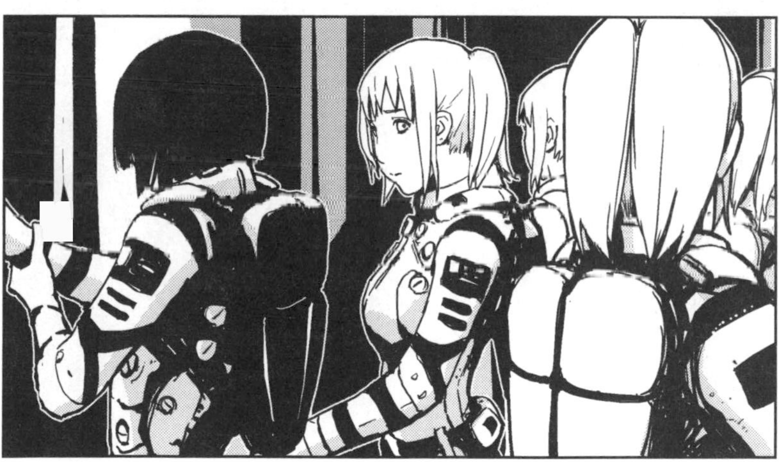

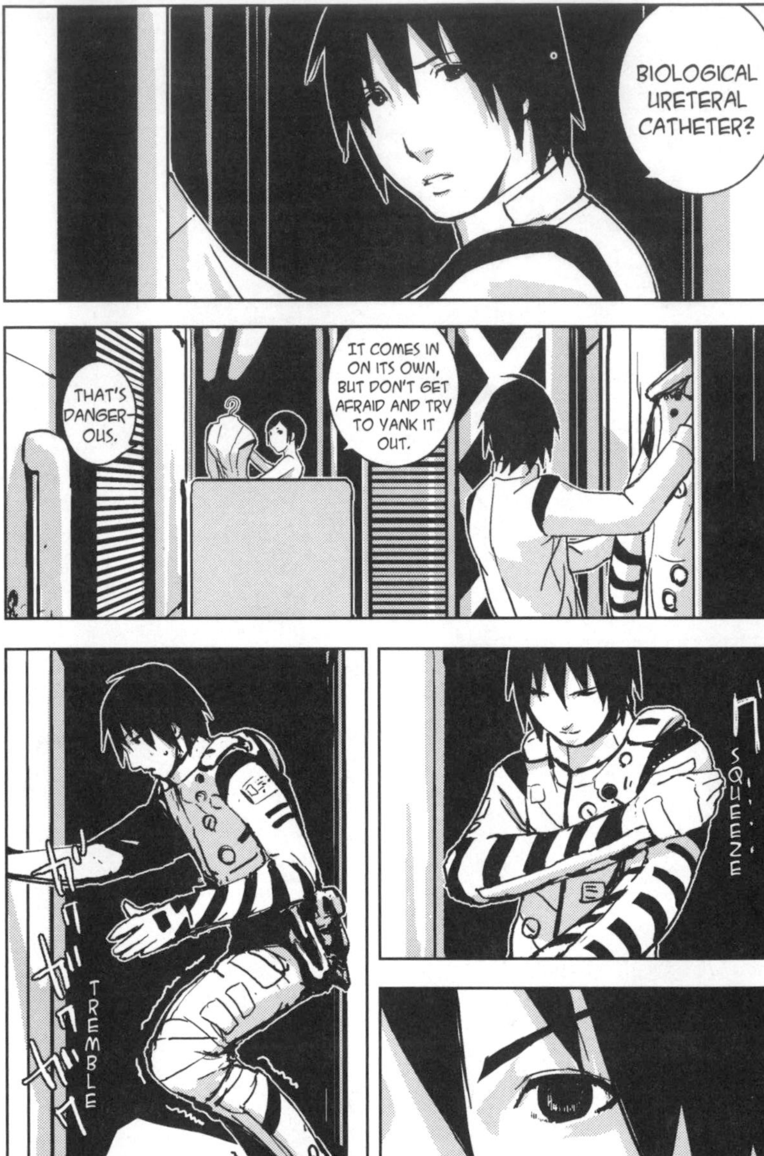
BIOLOGICAL URETERAL CATHETER?
IT COMES IN ON ITS OWN, BUT DON'T GET AFRAID AND TRY TO YANK IT OUT.
THAT'S DANGER-OUS.
SQUEEZE
TREMBLE

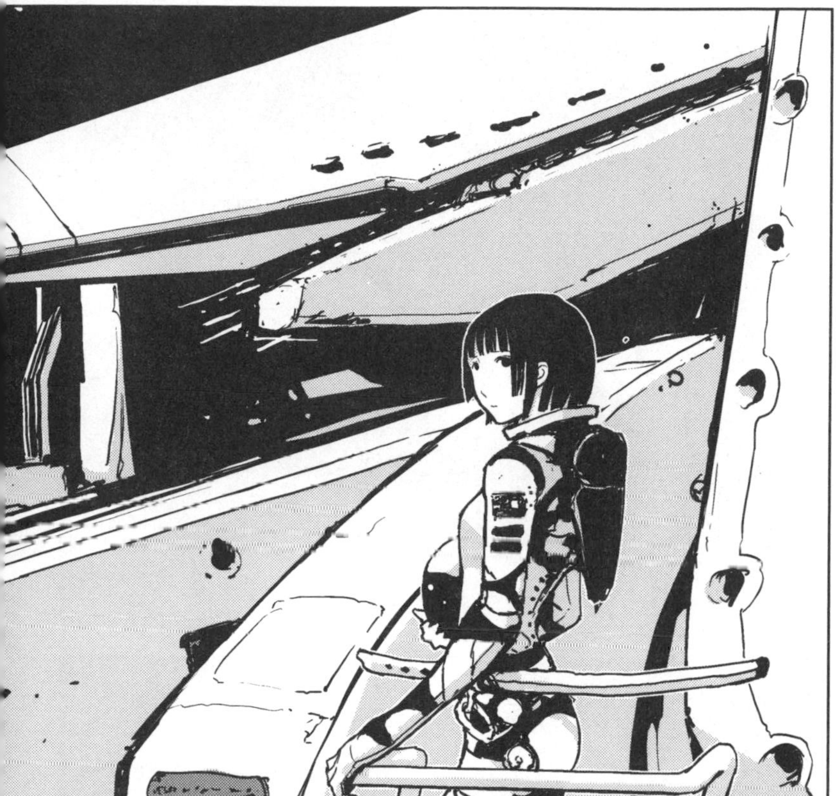

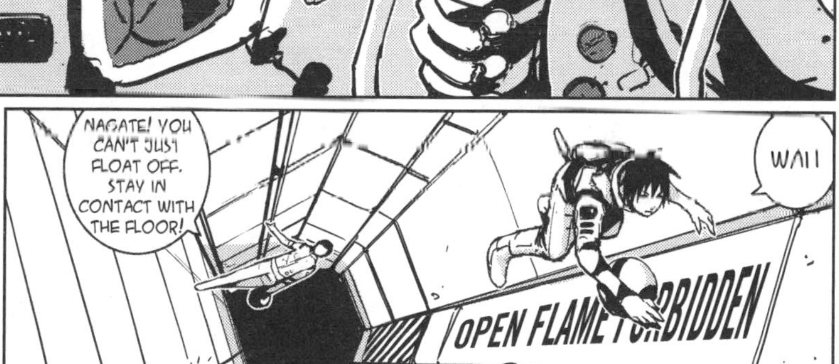
WAIT
NAGATE! YOU CAN'T JUST FLOAT OFF. STAY IN CONTACT WITH THE FLOOR!
OPEN FLAME FORBIDDEN

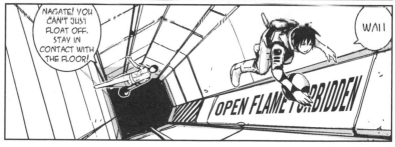

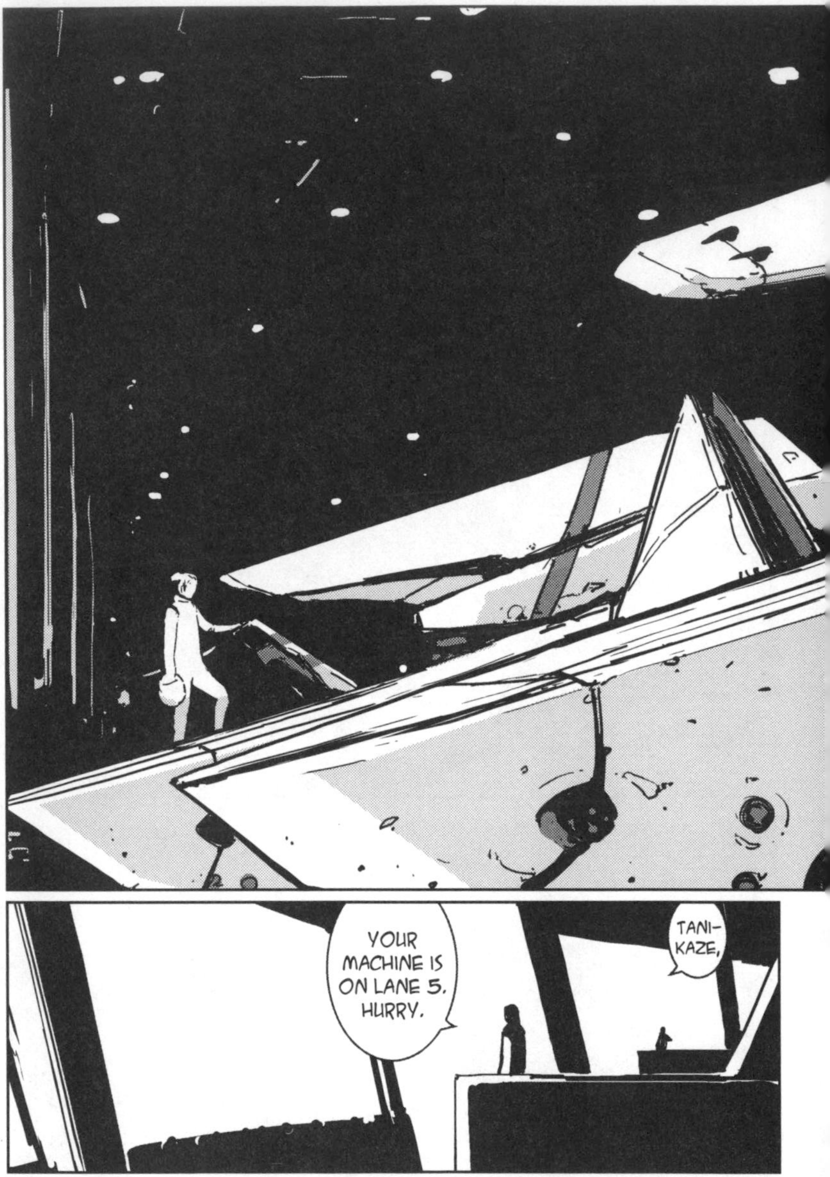
TANI-KAZE,
YOUR MACHINE IS ON LANE 5. HURRY.

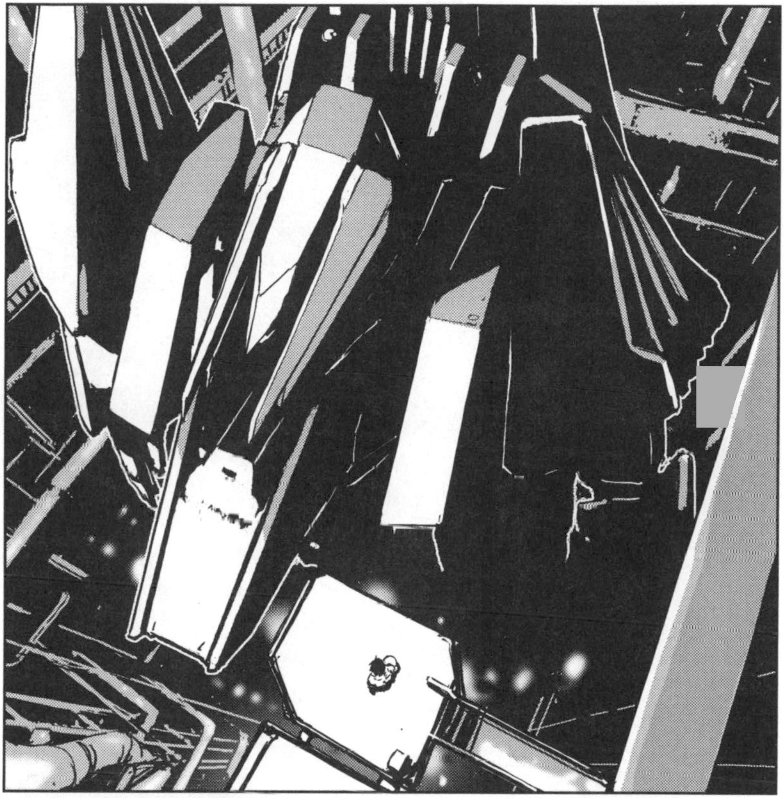

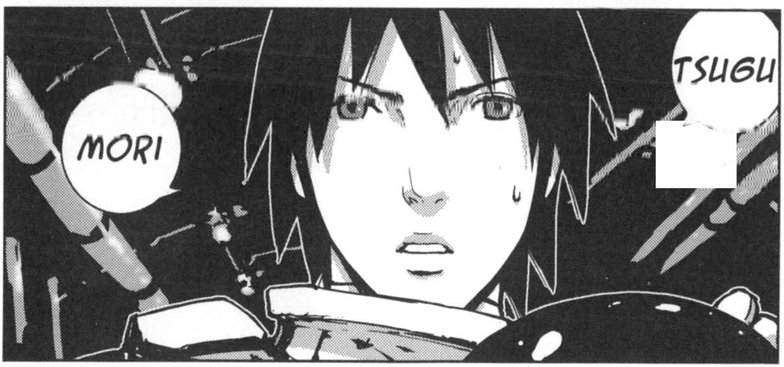
TSUGU
MORI

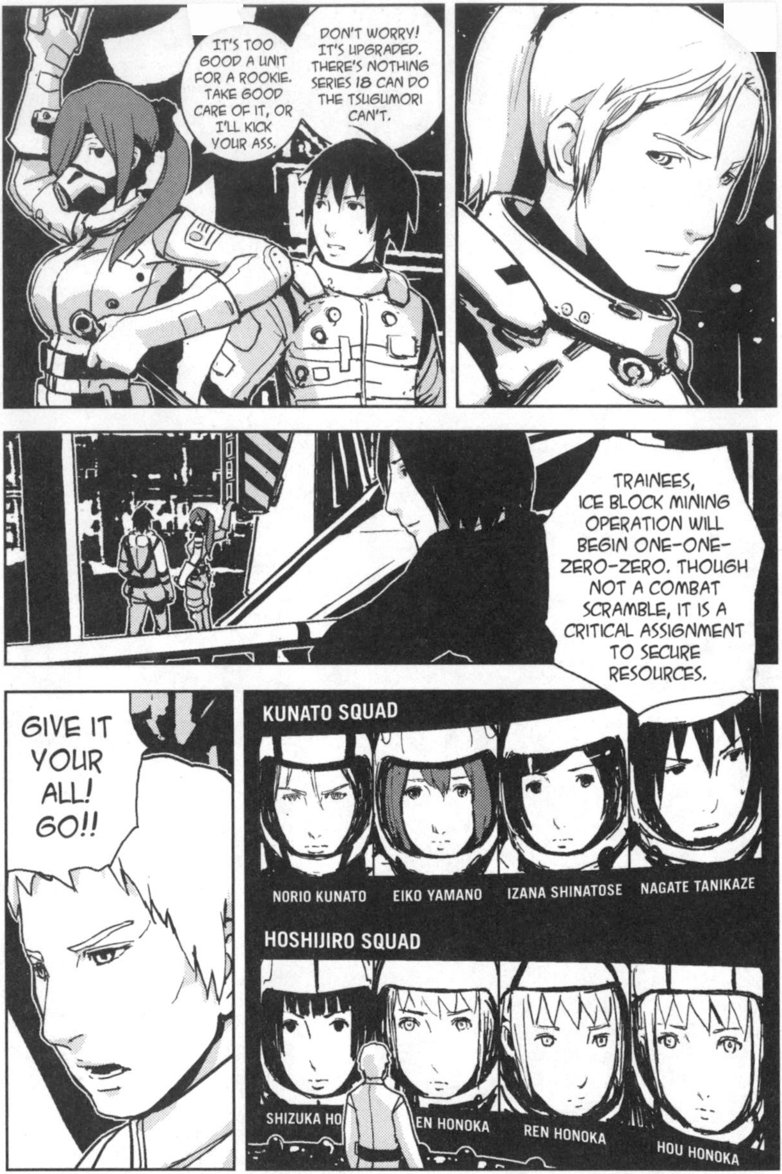
DON'T WORRY! IT'S UPGRADED. THERE'S NOTHING SERIES 18 CAN DO THE TSUGUMORI CAN'T.
IT'S TOO GOOD A UNIT FOR A ROOKIE. TAKE GOOD CARE OF IT, OR I'LL KICK YOUR ASS.
TRAINEES, ICE BLOCK MINING OPERATION WILL BEGIN ONE-ONE-ZERO-ZERO. THOUGH NOT A COMBAT SCRAMBLE, IT IS A CRITICAL ASSIGNMENT TO SECURE RESOURCES.
KUNATO SQUAD
NORIO KUNATO
EIKO YAMANO
IZANA SHINATOSE
NAGATE TANIKAZE
HOSHIJIRO SQUAD
SHIZUKA HO
EN HONOKA
REN HONOKA
HOU HONOKA
GIVE IT YOUR ALL! GO!!

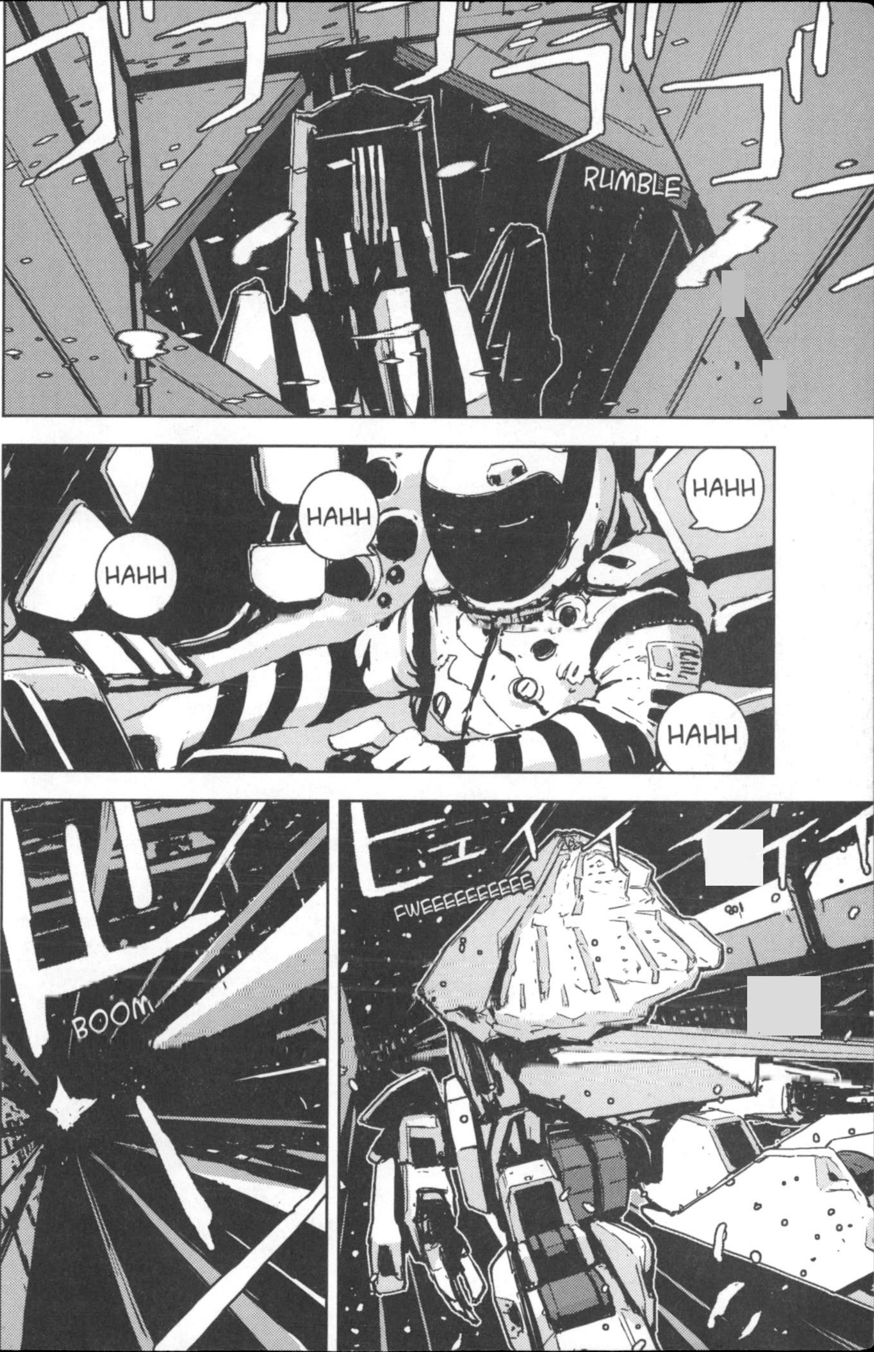
RUMBLE
HAHH
HAHH
HAHH
HAHH
FWEEEEEEEEEE
BOOM

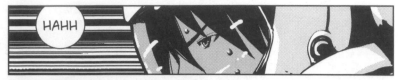

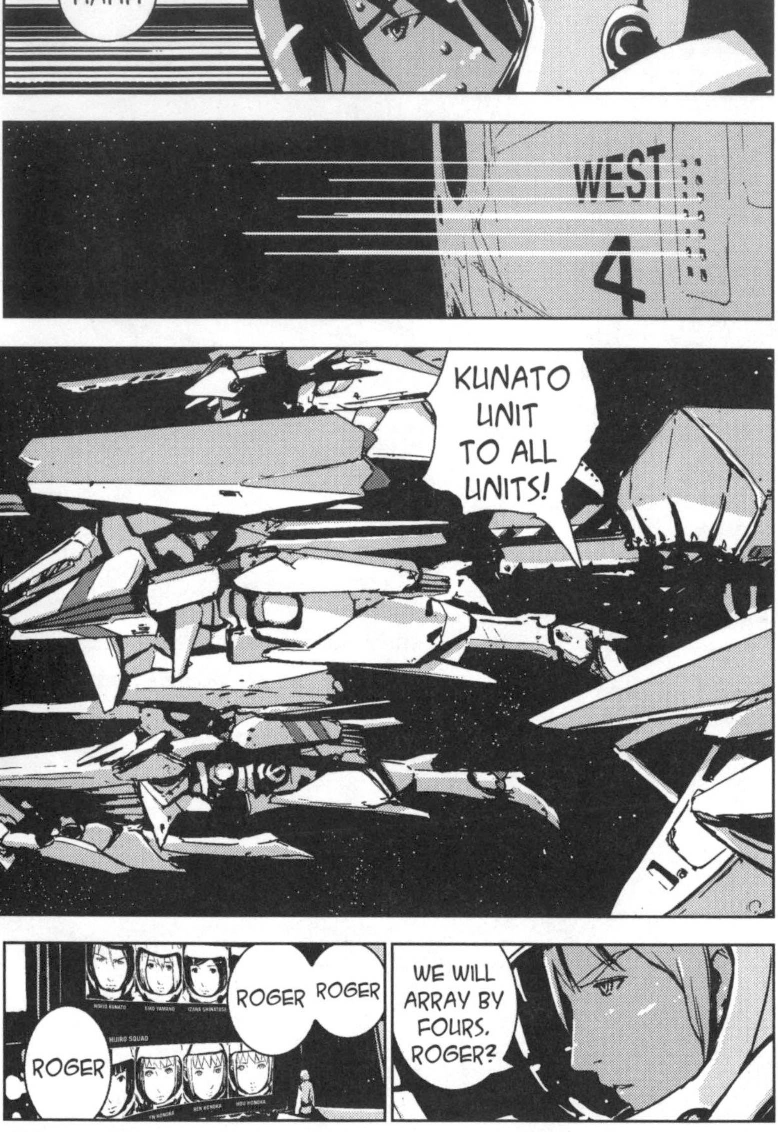
HAHH
WEST
4
KUNATO UNIT TO ALL UNITS!
1
WE WILL ARRAY BY FOURS. ROGER?
ROGER
ROGER
ROGER
NORIO KUNATO
EIKO YAMANO
IZANA SHINATOSE
HIJIRO SQUAD
HOU HONOKA
REN HONOKA

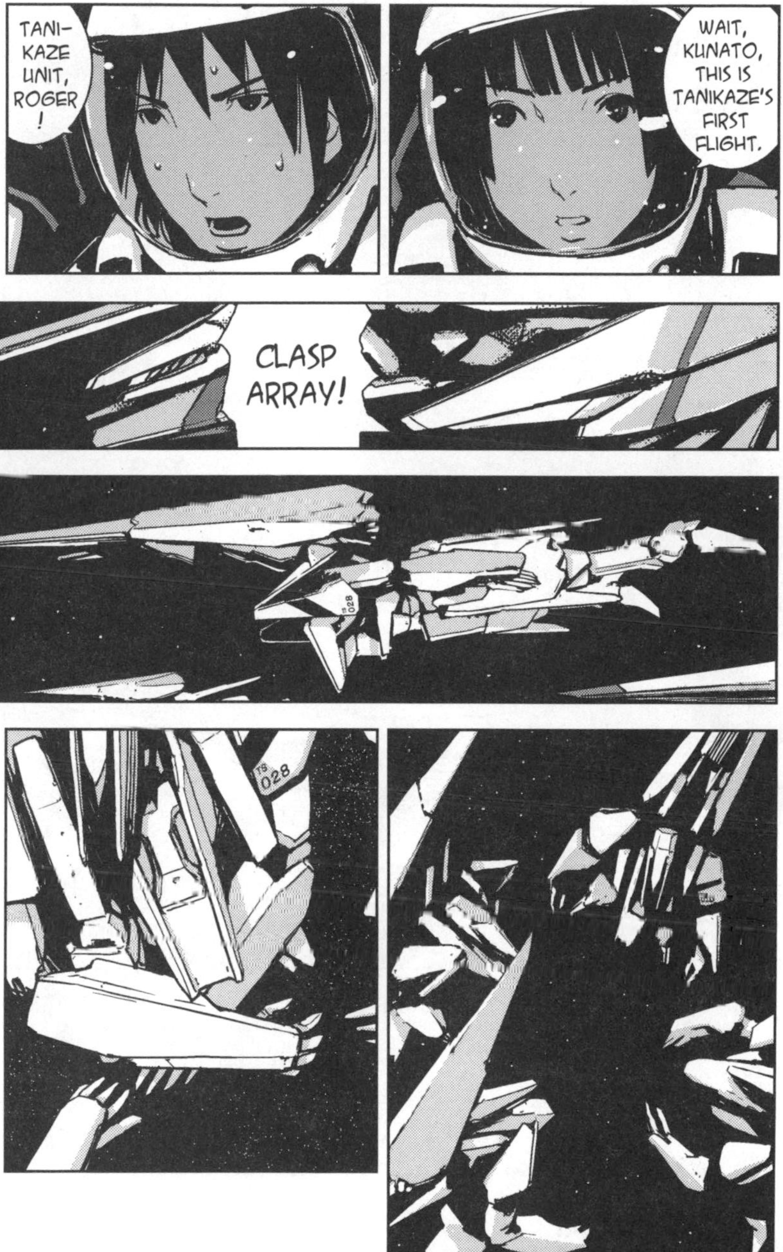
WAIT, KUNATO, THIS IS TANIKAZE'S FIRST FLIGHT.
TANI-KAZE UNIT, ROGER!
CLASP ARRAY!

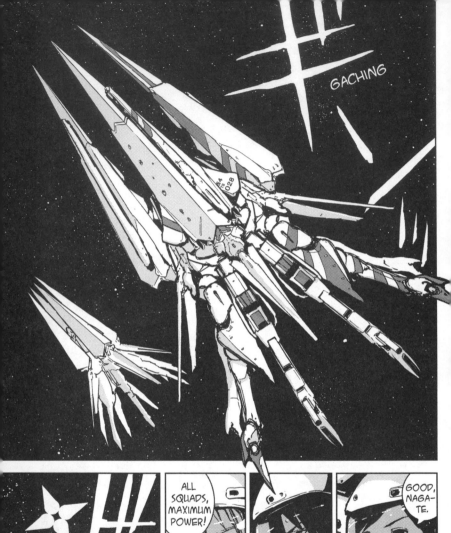

GACHING

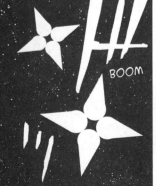

BOOM

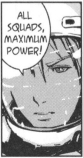

ALL SQUADS, MAXIMUM POWER!

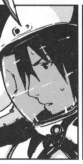

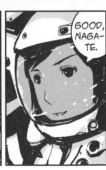

GOOD, NAGATE.

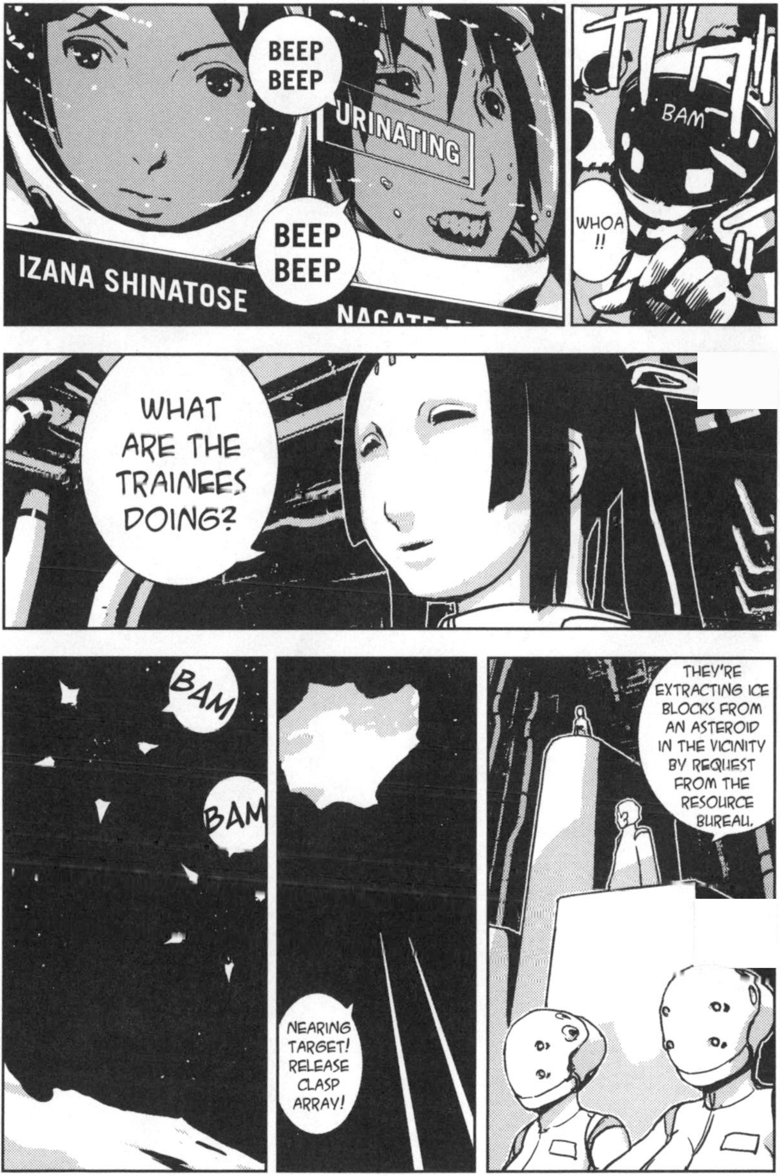
BAM
WHOA!!
BEEP BEEP
URINATING
BEEP BEEP
IZANA SHINATOSE
NAGATE
WHAT ARE THE TRAINEES DOING?
THEY'RE EXTRACTING ICE BLOCKS FROM AN ASTEROID IN THE VICINITY BY REQUEST FROM THE RESOURCE BUREAU.
NEARING TARGET! RELEASE CLASP ARRAY!
BAM
BAM

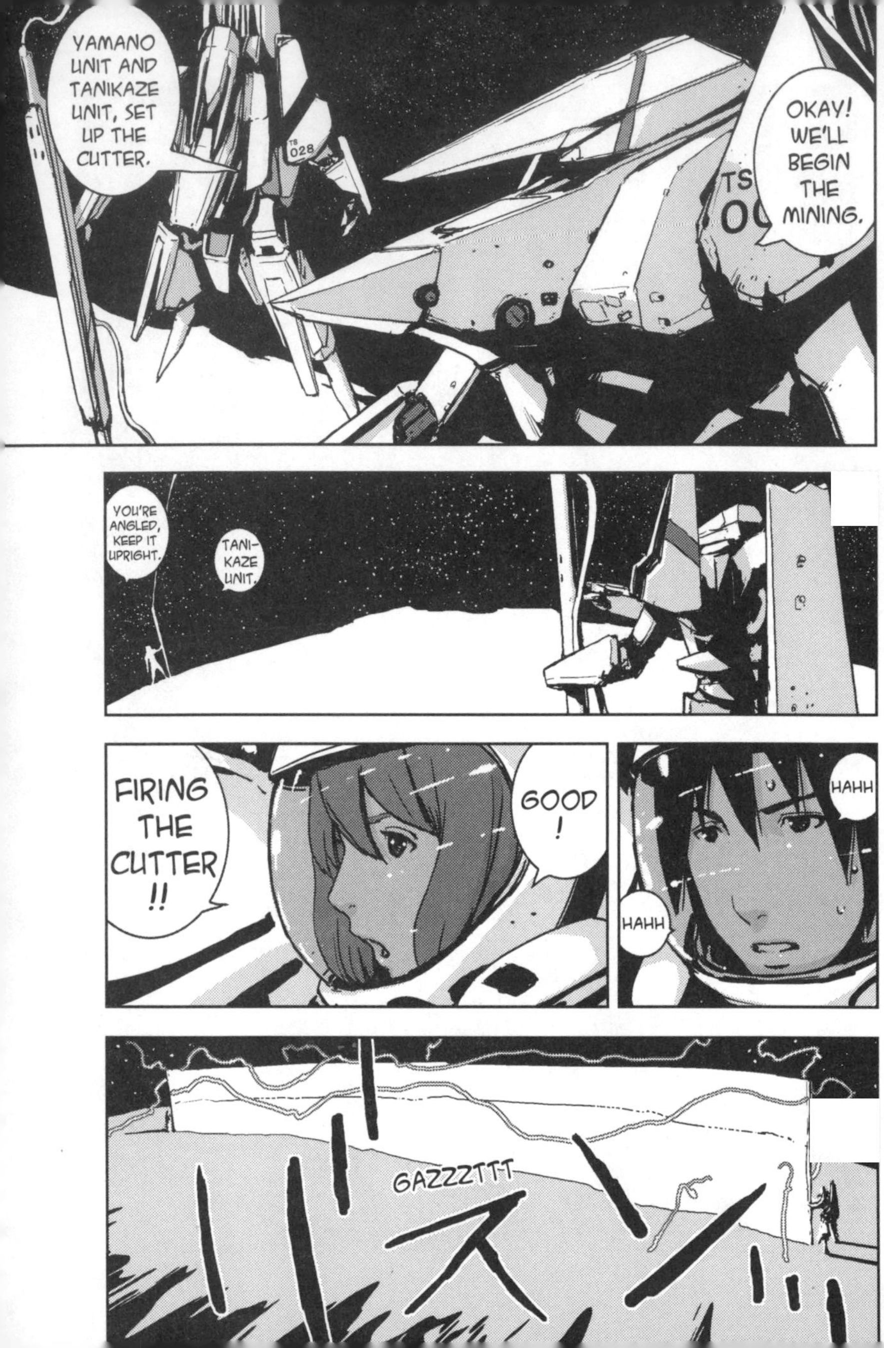
OKAY! WE'LL BEGIN THE MINING.
YAMANO UNIT AND TANIKAZE UNIT, SET UP THE CUTTER.
TANI-KAZE UNIT.
YOU'RE ANGLED, KEEP IT UPRIGHT.
HAHH
HAHH
GOOD!
FIRING THE CUTTER!!
GAZZZTTT

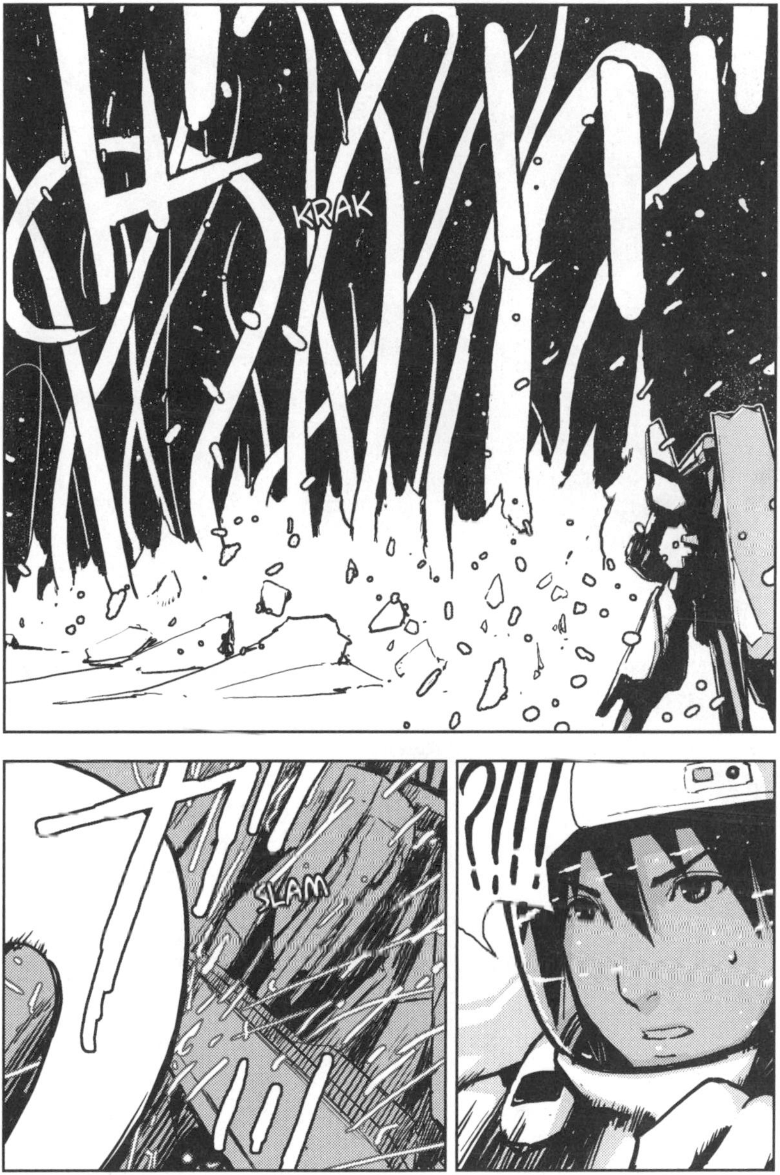
KRAK
?!!
SLAM

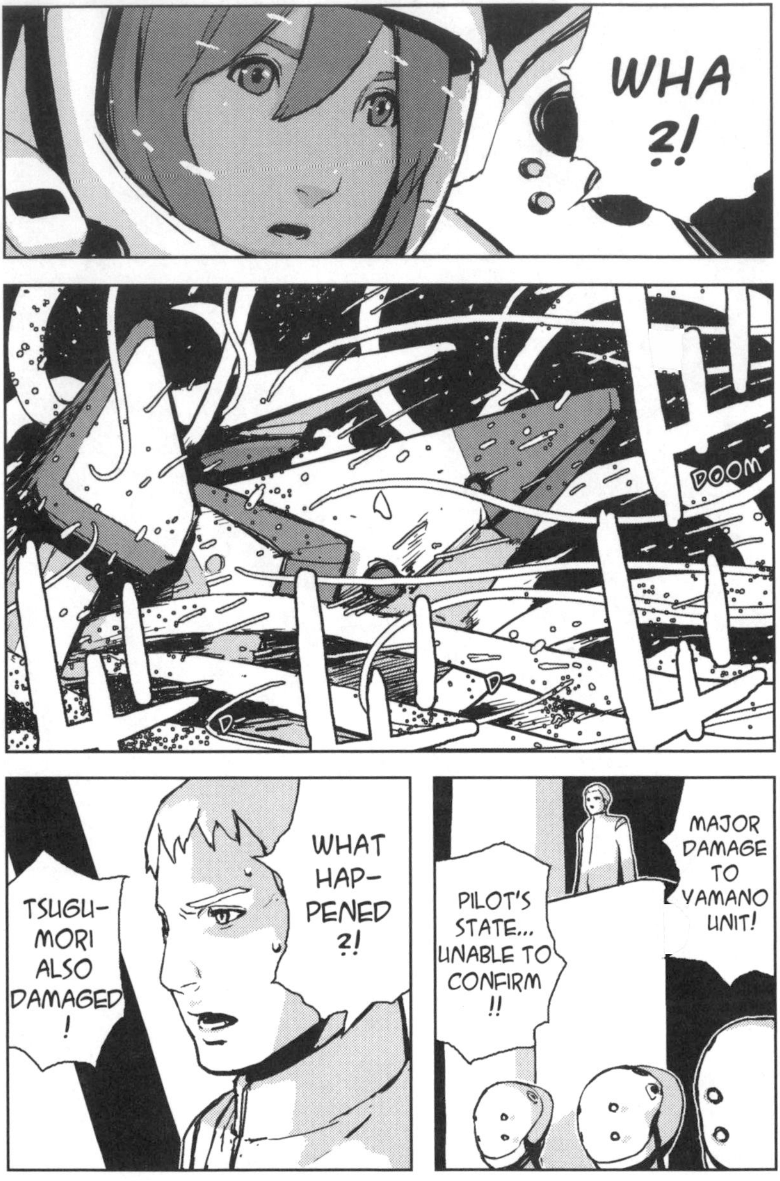
WHA ?!
DOOM
MAJOR DAMAGE TO YAMANO UNIT!
PILOT'S STATE... UNABLE TO CONFIRM !!
WHAT HAP-PENED ?!
TSUGU-MORI ALSO DAMAGED !

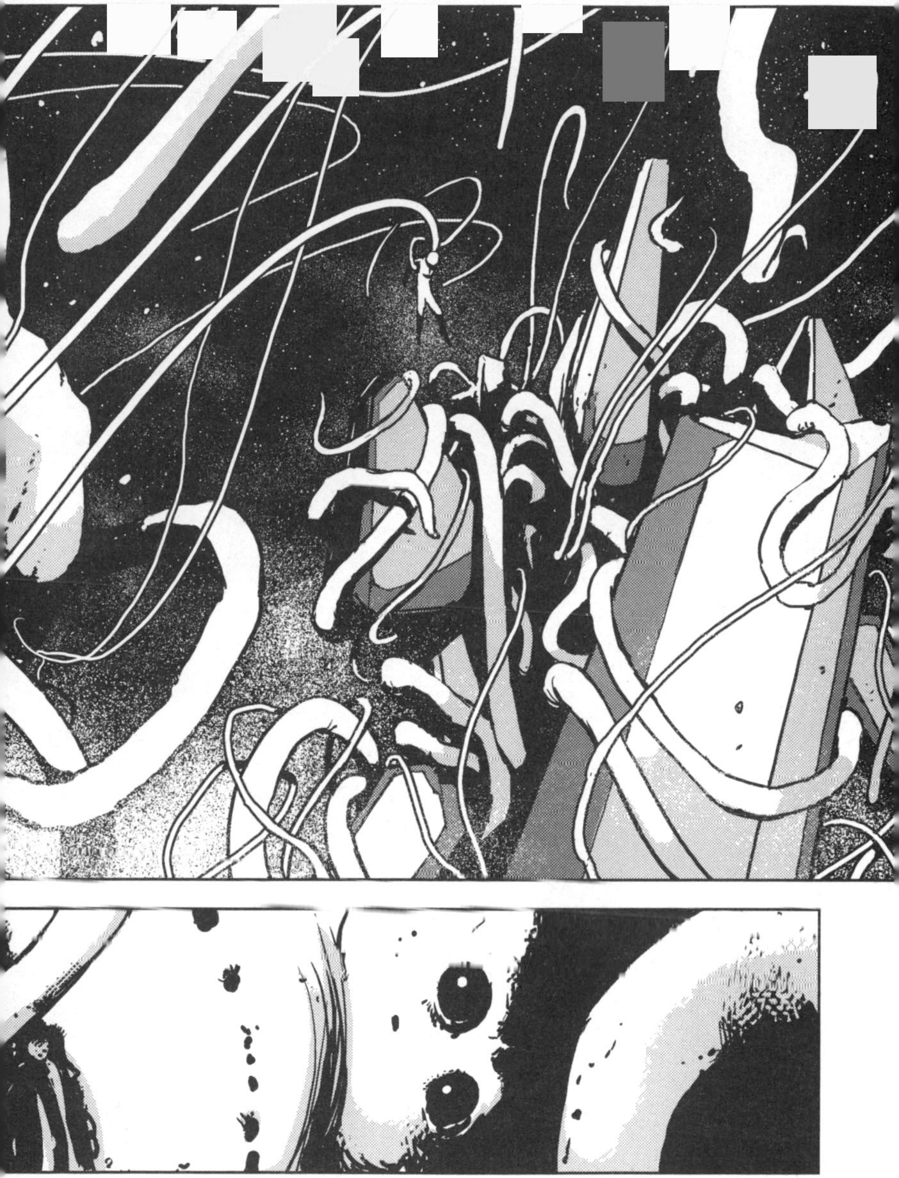

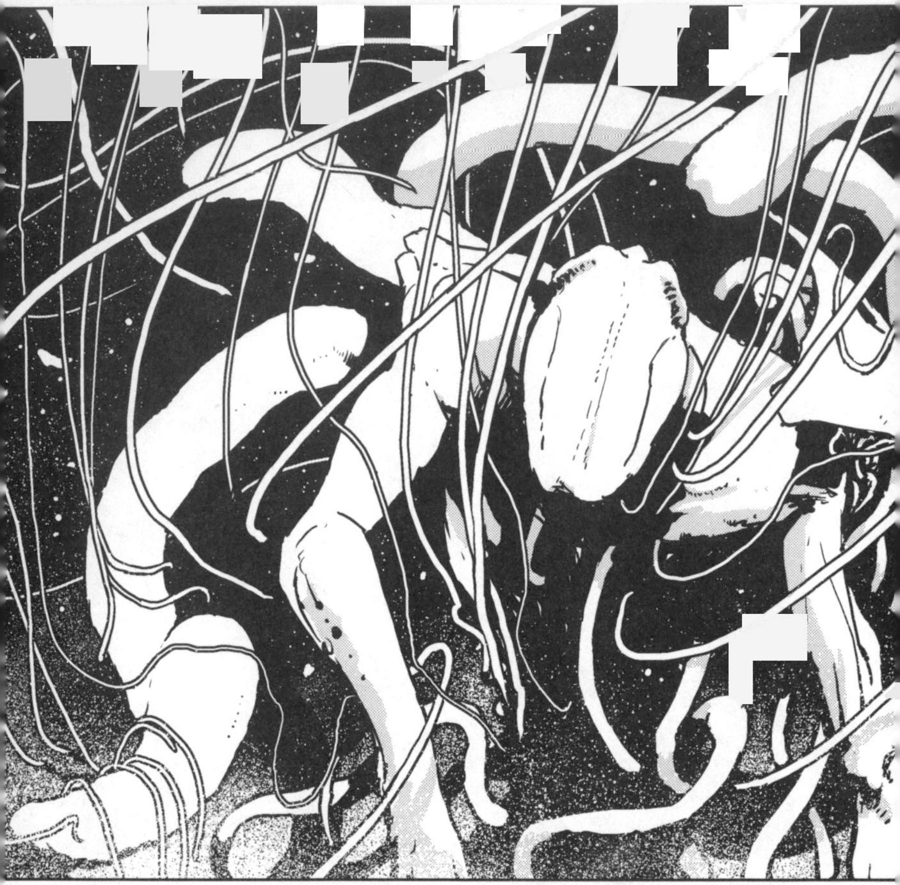

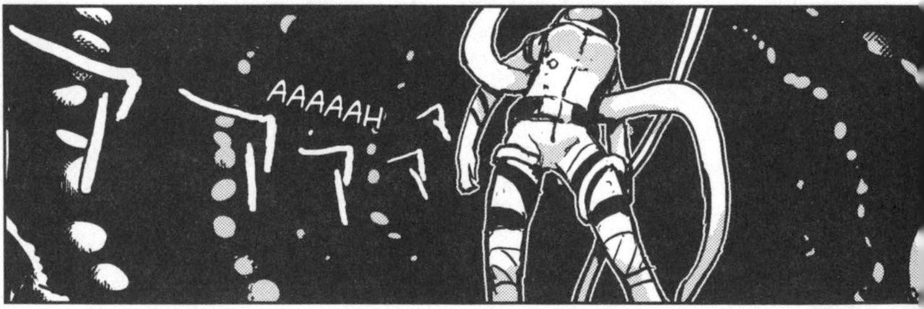
AAAAAH!

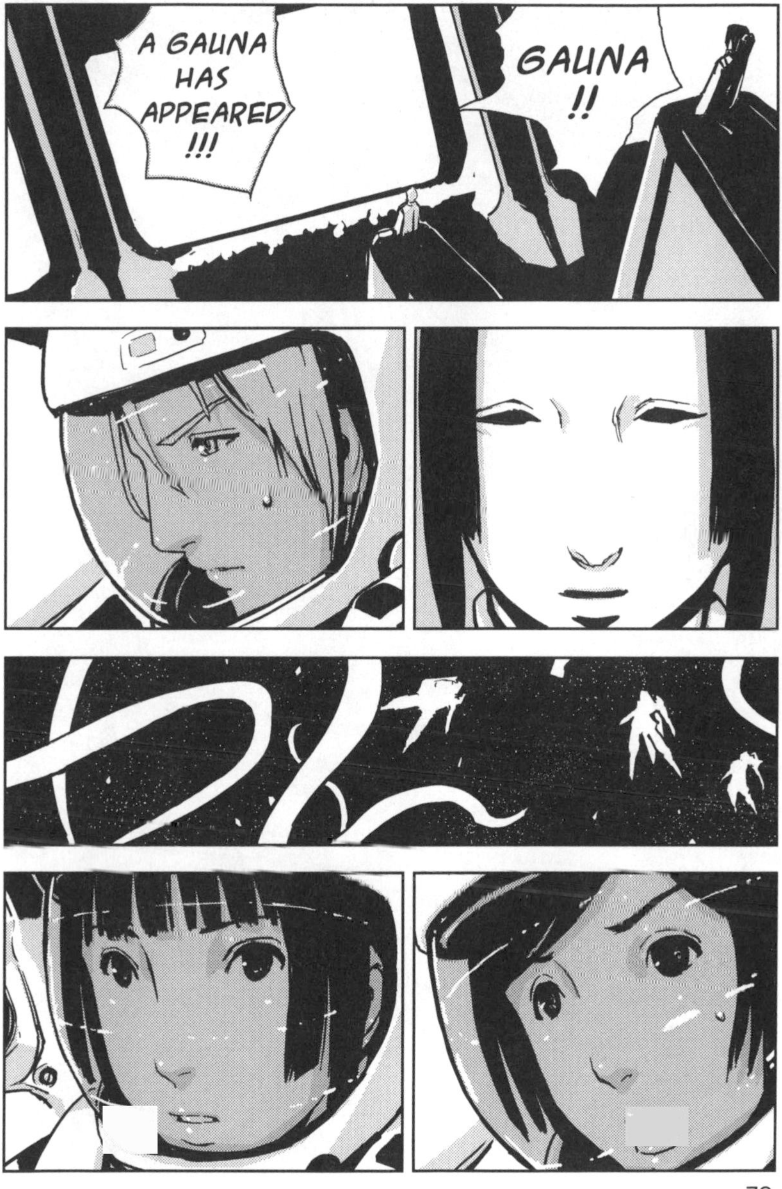
GAUNA !!
A GAUNA HAS APPEARED !!!

BEEP
BEEP
BEEP
NAGATE TANIKAZE
CARDIOPULMONARY ARREST

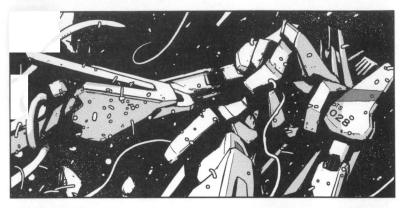

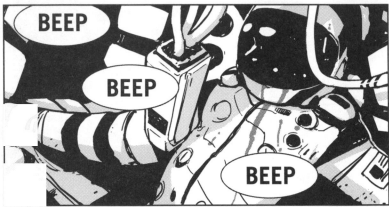

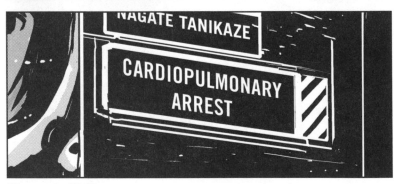

Chapter 2: END

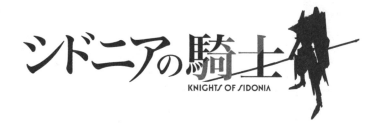

シドニアの騎士
KNIGHTS OF SIDONIA

Chapter 3: Eiko Yamano's Starry Heavens

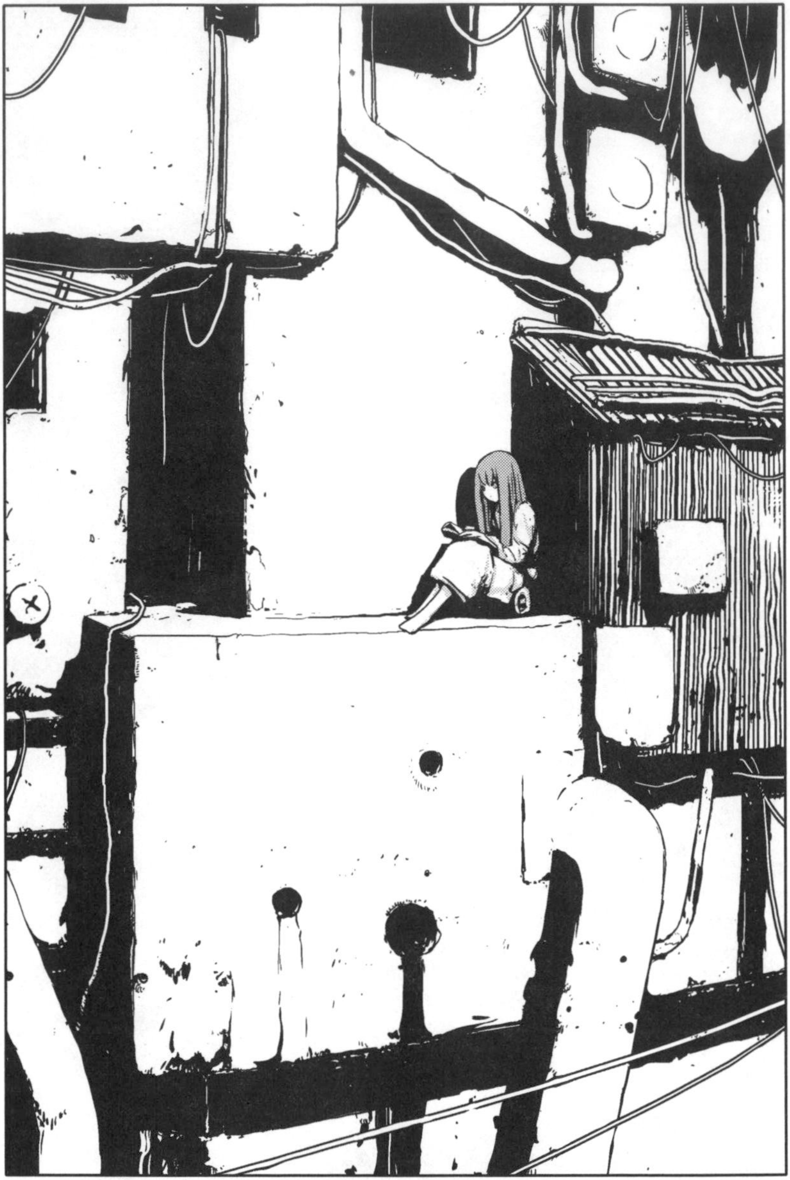

One Hundred Sights of Sidonia Part Two:
Sublevel Unstable Motive Power Area

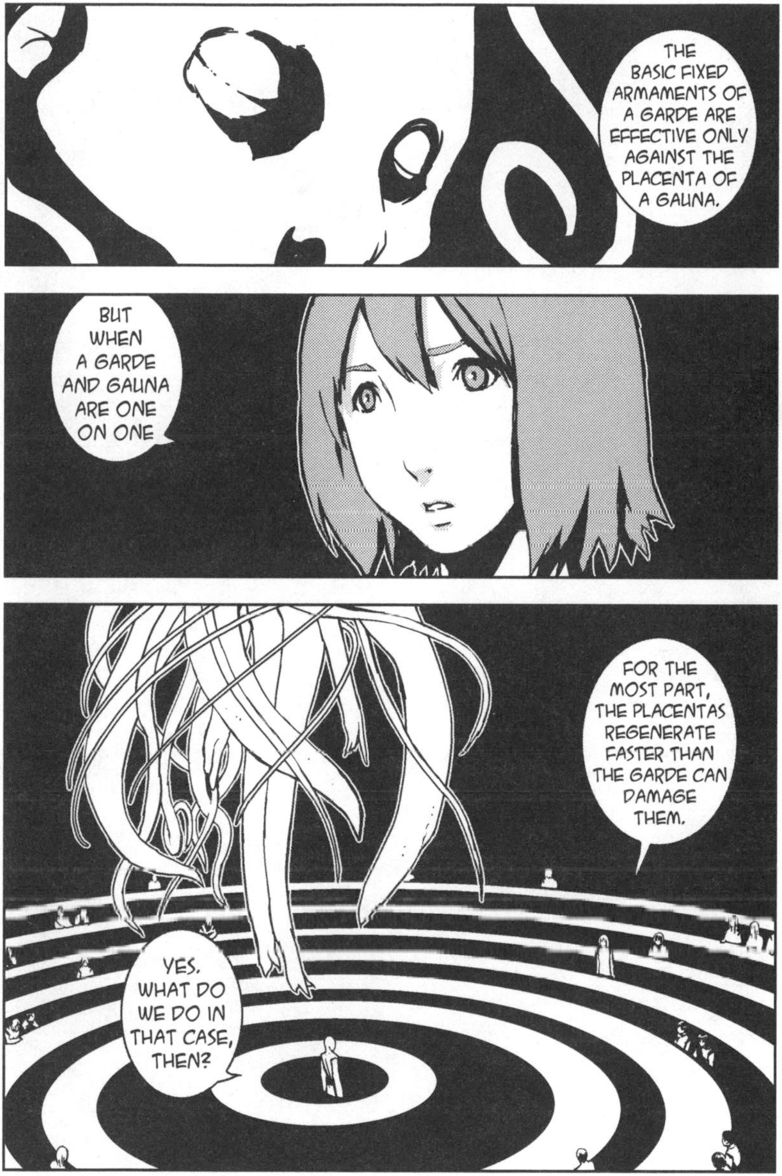
THE BASIC FIXED ARMAMENTS OF A GARDE ARE EFFECTIVE ONLY AGAINST THE PLACENTA OF A GAUNA.
BUT WHEN A GARDE AND GAUNA ARE ONE ON ONE
FOR THE MOST PART, THE PLACENTAS REGENERATE FASTER THAN THE GARDE CAN DAMAGE THEM.
YES. WHAT DO WE DO IN THAT CASE, THEN?

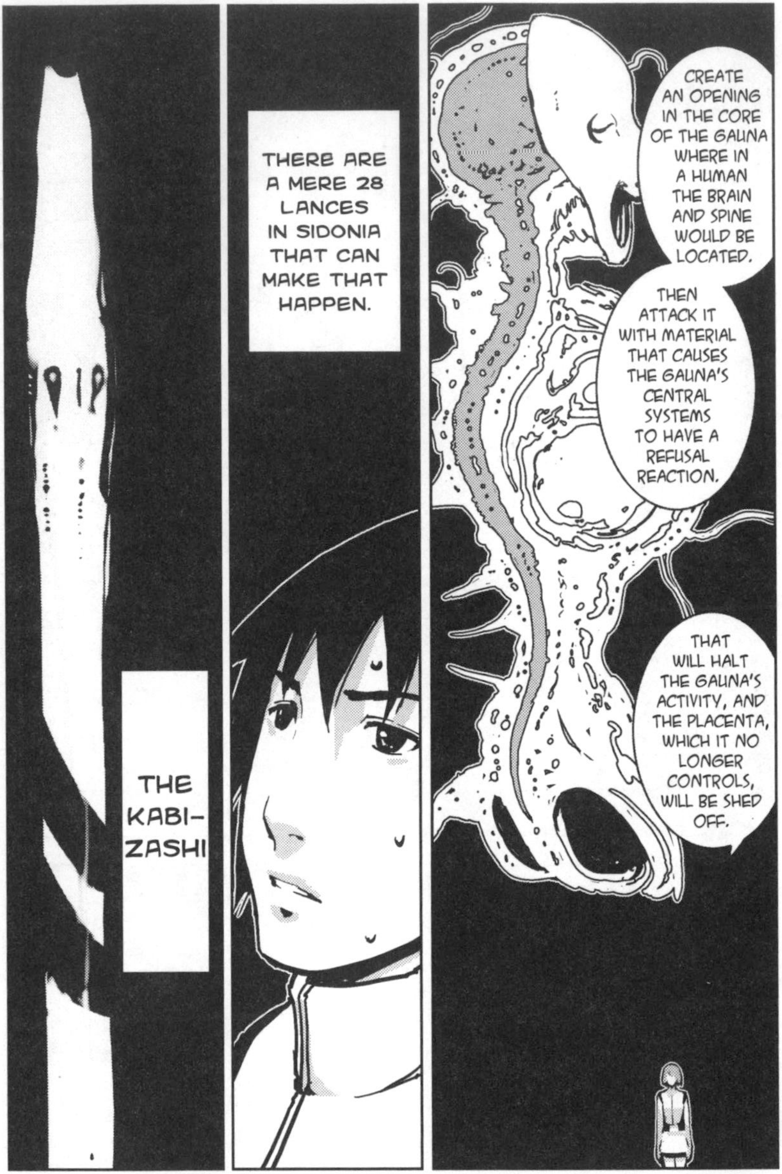
CREATE AN OPENING IN THE CORE OF THE GAUNA WHERE IN A HUMAN THE BRAIN AND SPINE WOULD BE LOCATED.
THEN ATTACK IT WITH MATERIAL THAT CAUSES THE GAUNA'S CENTRAL SYSTEMS TO HAVE A REFUSAL REACTION.
THAT WILL HALT THE GAUNA'S ACTIVITY, AND THE PLACENTA, WHICH IT NO LONGER CONTROLS, WILL BE SHED OFF.
THERE ARE A MERE 28 LANCES IN SIDONIA THAT CAN MAKE THAT HAPPEN.
THE KABI-ZASHI

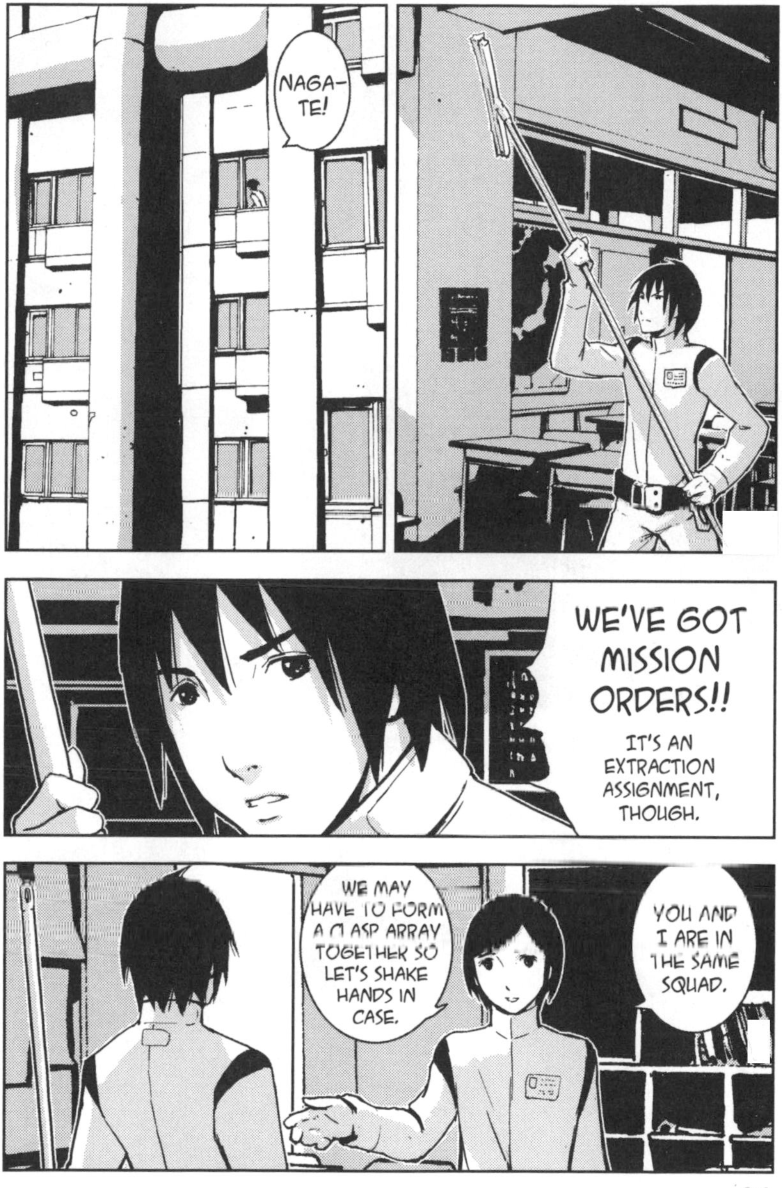
NAGA-TE!
WE'VE GOT MISSION ORDERS!!
IT'S AN EXTRACTION ASSIGNMENT, THOUGH.
YOU AND I ARE IN THE SAME SQUAD.
WE MAY HAVE TO FORM A CLASP ARRAY TOGETHER SO LET'S SHAKE HANDS IN CASE.

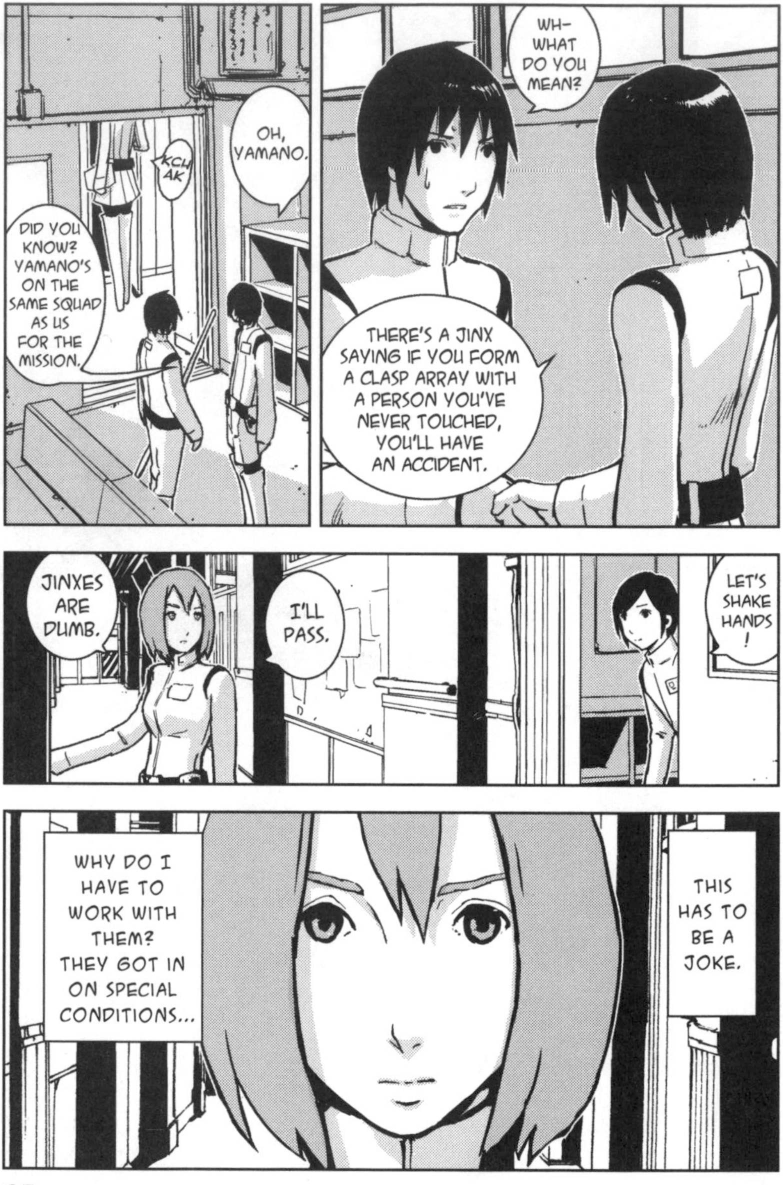
WH- WHAT DO YOU MEAN?
THERE'S A JINX SAYING IF YOU FORM A CLASP ARRAY WITH A PERSON YOU'VE NEVER TOUCHED, YOU'LL HAVE AN ACCIDENT.
OH, YAMANO.
KCHAK
DID YOU KNOW? YAMANO'S ON THE SAME SQUAD AS US FOR THE MISSION.
LET'S SHAKE HANDS!
I'LL PASS.
JINXES ARE DUMB.
THIS HAS TO BE A JOKE.
WHY DO I HAVE TO WORK WITH THEM? THEY GOT IN ON SPECIAL CONDITIONS...

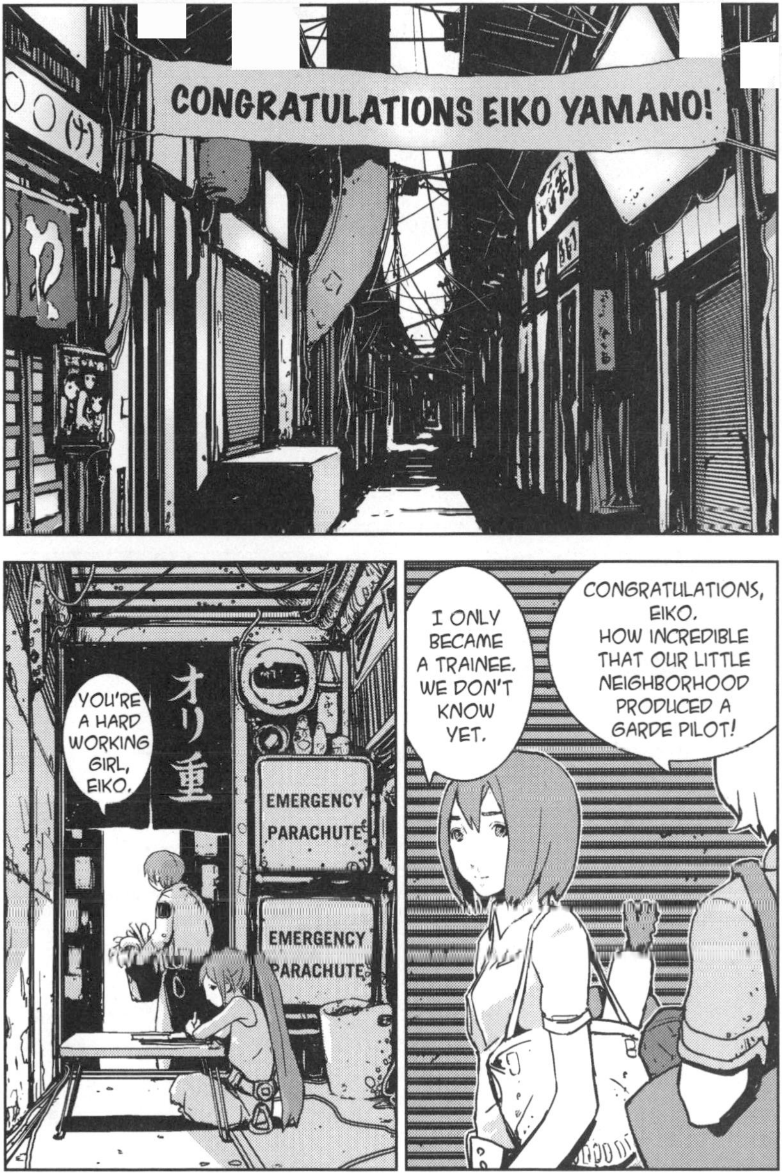
CONGRATULATIONS EIKO YAMANO!
CONGRATULATIONS, EIKO. HOW INCREDIBLE THAT OUR LITTLE NEIGHBORHOOD PRODUCED A GARDE PILOT!
I ONLY BECAME A TRAINEE. WE DON'T KNOW YET.
YOU'RE A HARD WORKING GIRL, EIKO.
オリ重
EMERGENCY PARACHUTE
EMERGENCY PARACHUTE

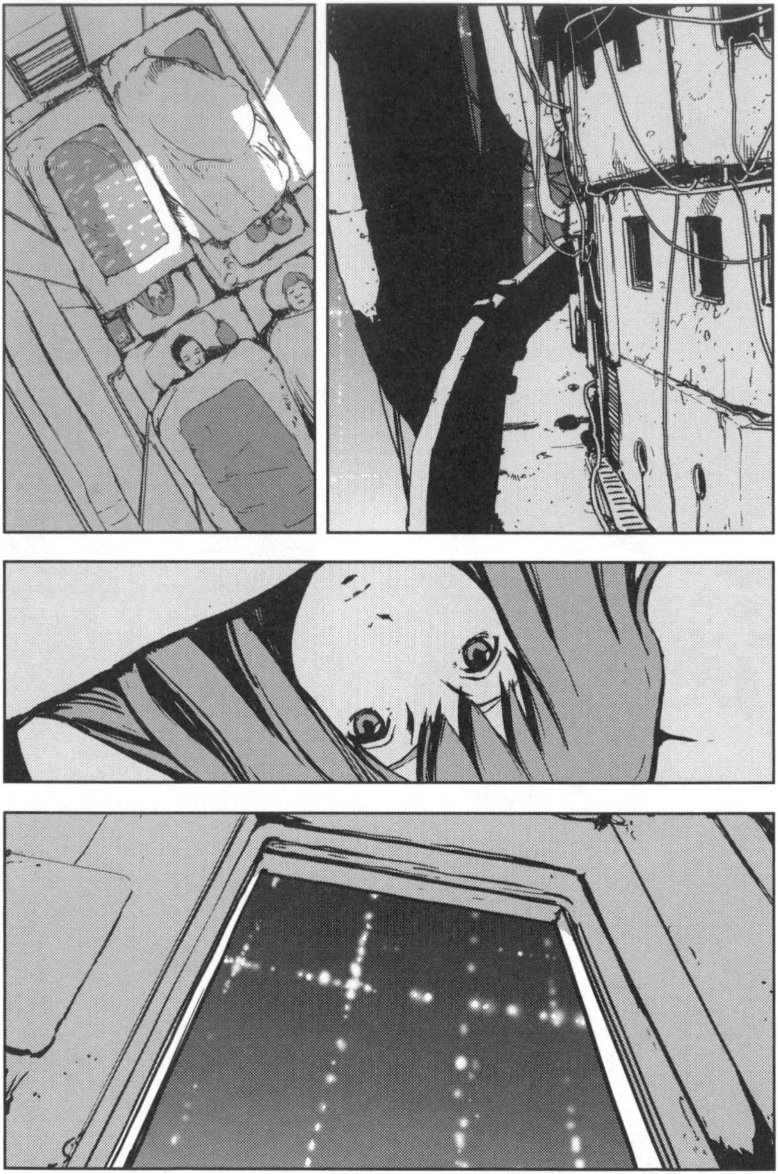

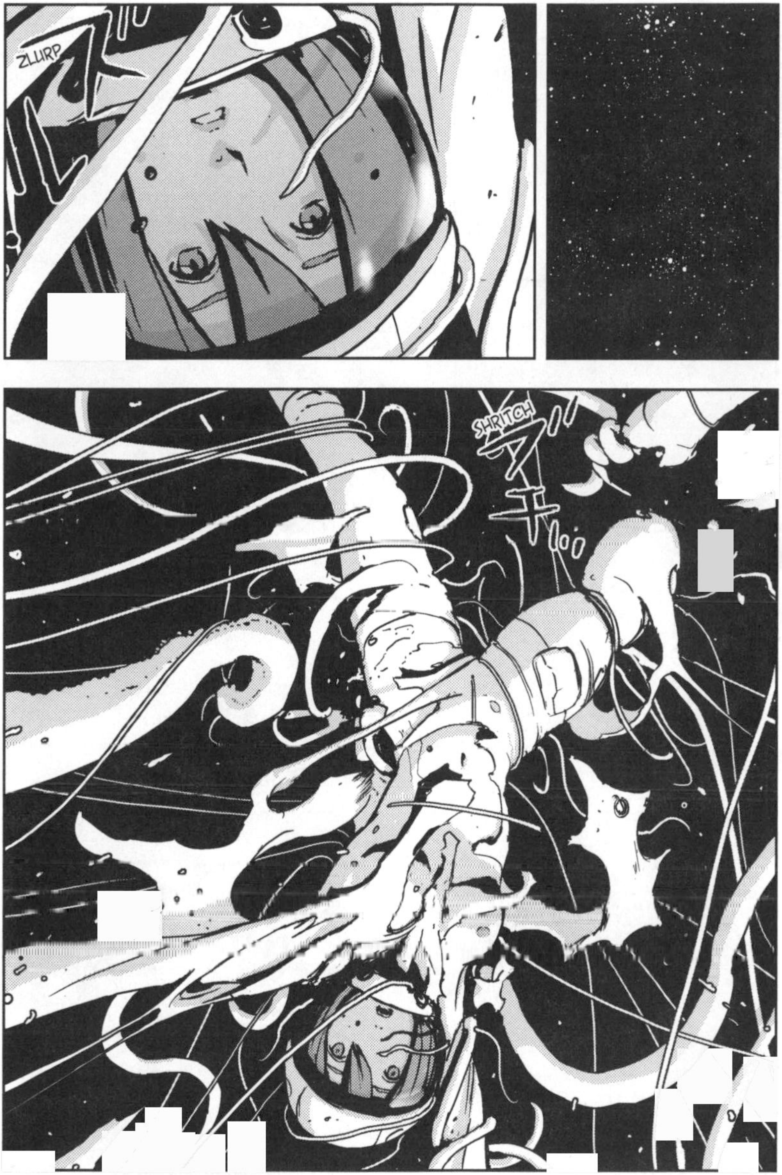
ZLURP
SHRITCH

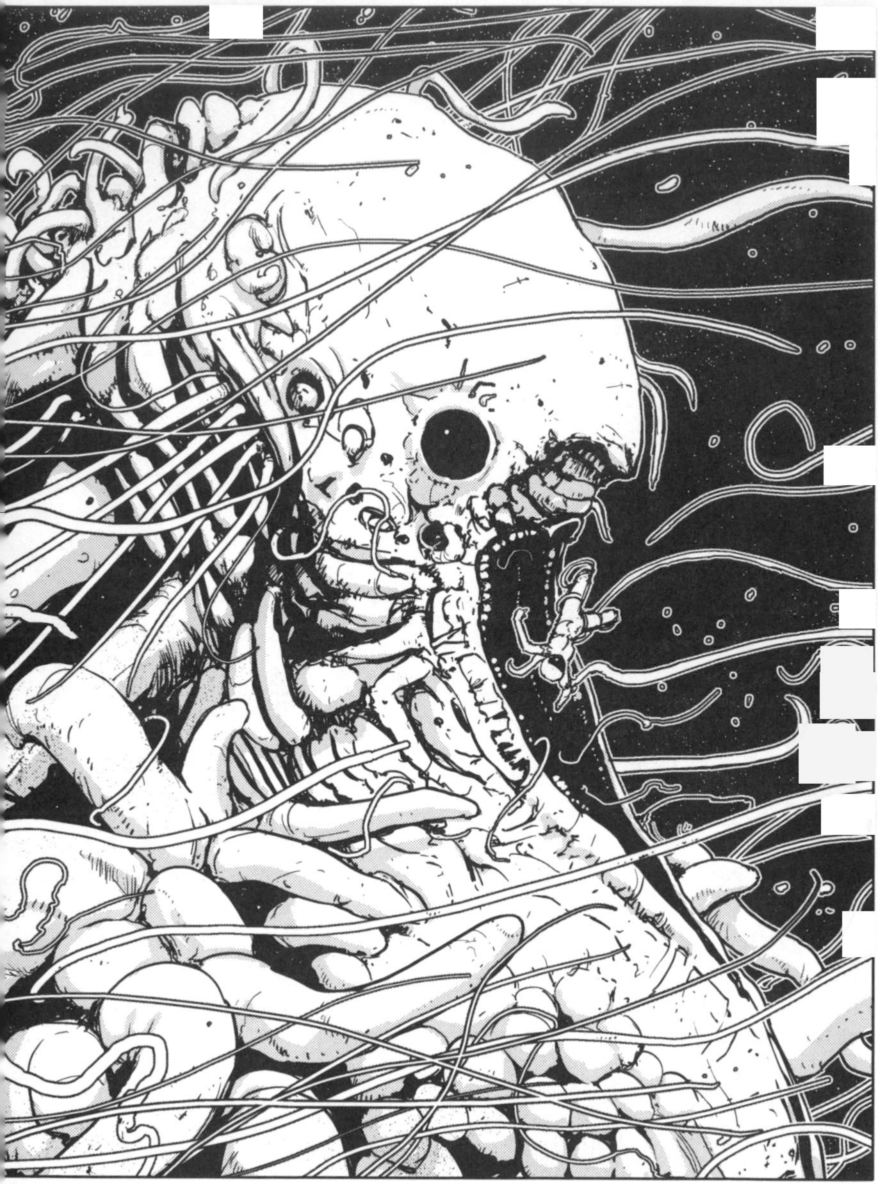

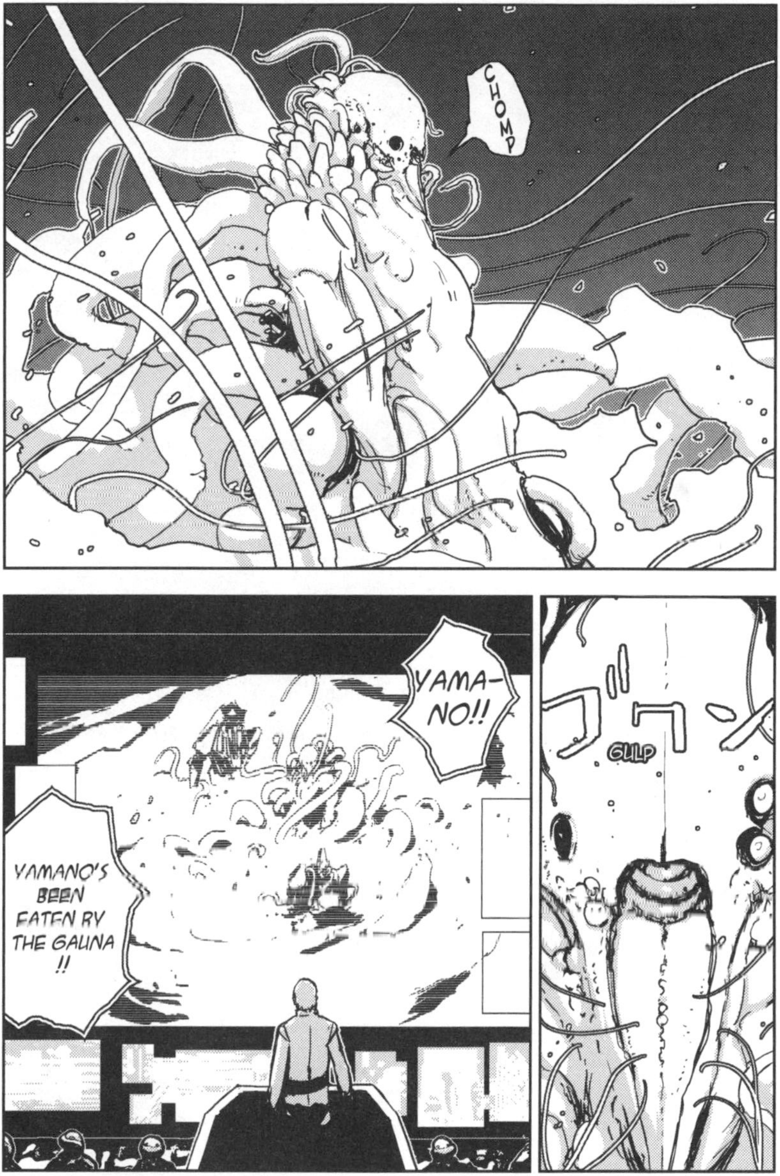
CHOMP
GULP
YAMANO!!
YAMANO'S BEEN EATEN BY THE GAUNA!!

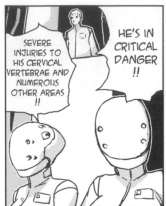

WHAT ABOUT TANIKAZE?!
WHAT'S TANIKAZE'S CONDITION?
HE'S IN CRITICAL DANGER!!
SEVERE INJURIES TO HIS CERVICAL VERTEBRAE AND NUMEROUS OTHER AREAS!!
HE ISN'T HEALABLE JUST WITH THE SUIT'S RESUSCITATION EQUIPMENT!
NAGATE TANIKAZE
CARDIOPULMONARY ARREST
HIS UNIT IS HELD FAST BY THE GAUNA!
IF WE FIRE A KABIZASHI FROM SIDONIA, WILL THE TRAINEES BE ABLE TO CATCH IT?
THEY HAVEN'T REACHED THAT STAGE OF TRAINING YET.
...

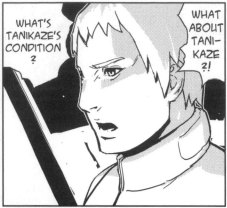

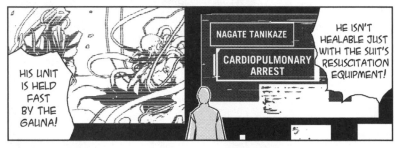

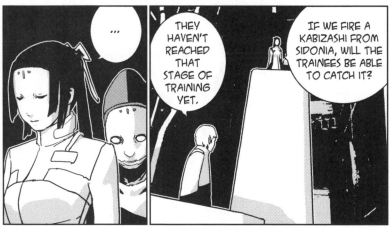

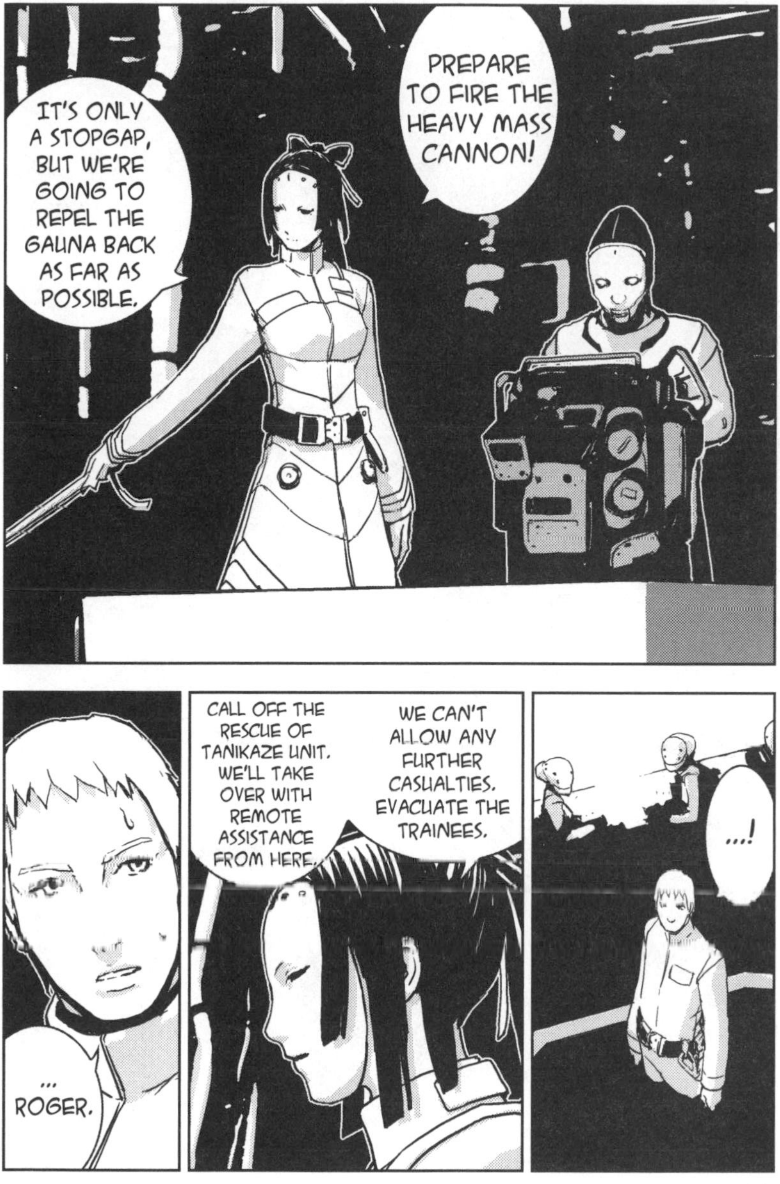
PREPARE TO FIRE THE HEAVY MASS CANNON!
IT'S ONLY A STOPGAP, BUT WE'RE GOING TO REPEL THE GAUNA BACK AS FAR AS POSSIBLE.
...!
WE CAN'T ALLOW ANY FURTHER CASUALTIES. EVACUATE THE TRAINEES.
CALL OFF THE RESCUE OF TANIKAZE UNIT. WE'LL TAKE OVER WITH REMOTE ASSISTANCE FROM HERE.
... ROGER.

GRRRRR
KONG
PING
PING
FIRE !!

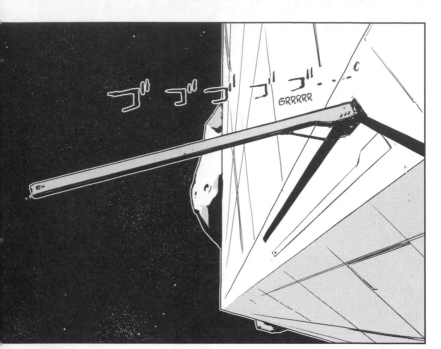

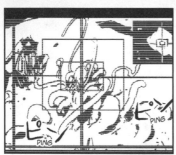

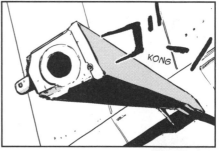

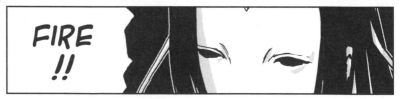

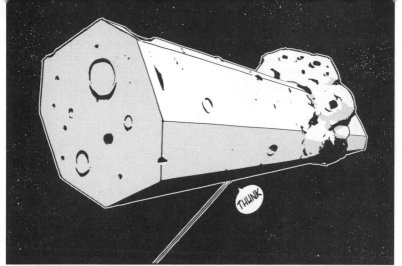

THUNK

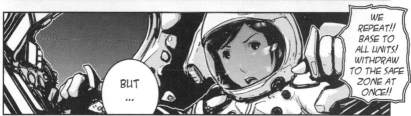

WE REPEAT!! BASE TO ALL UNITS! WITHDRAW TO THE SAFE ZONE AT ONCE!!

BUT ...

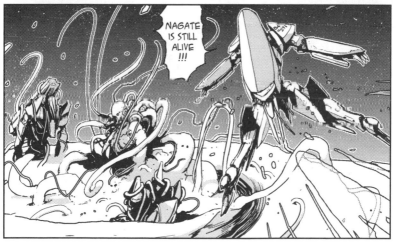

NAGATE IS STILL ALIVE !!!

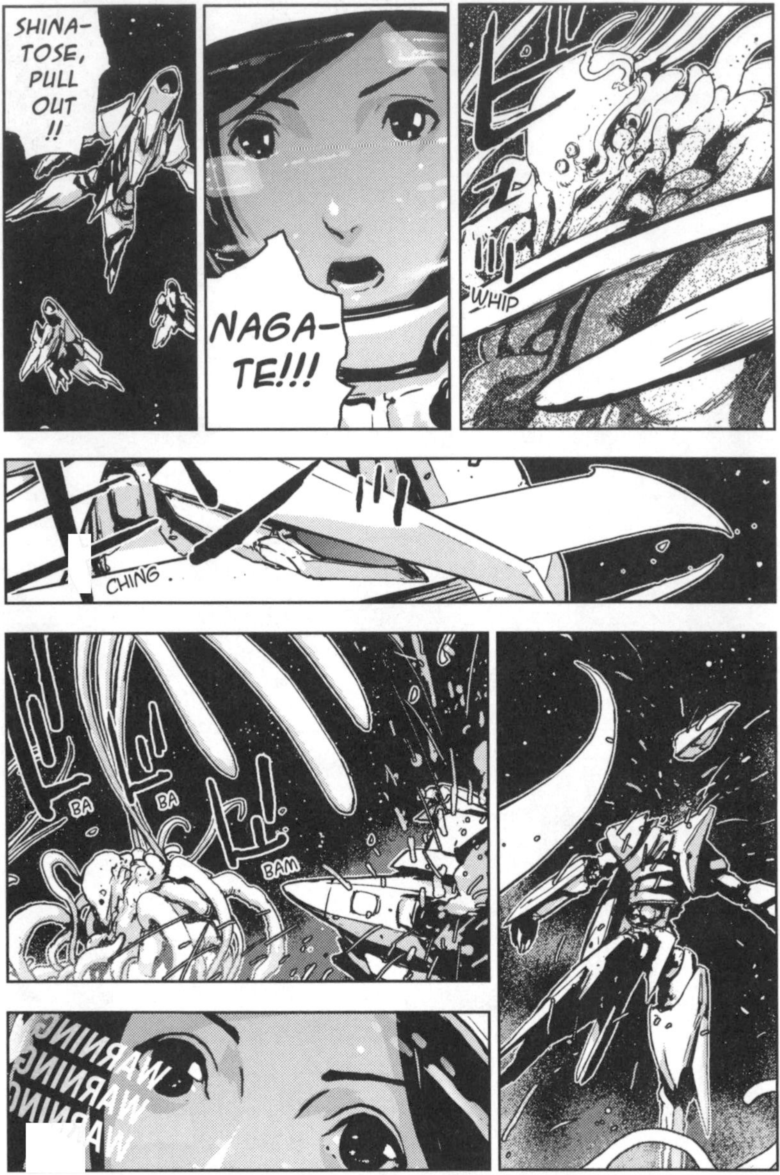
WHIP
NAGA-
TE!!!
SHINA-
TOSE,
PULL
OUT
!!
CHING
BA
BA
BAM
WARNING
WARNING
WARN

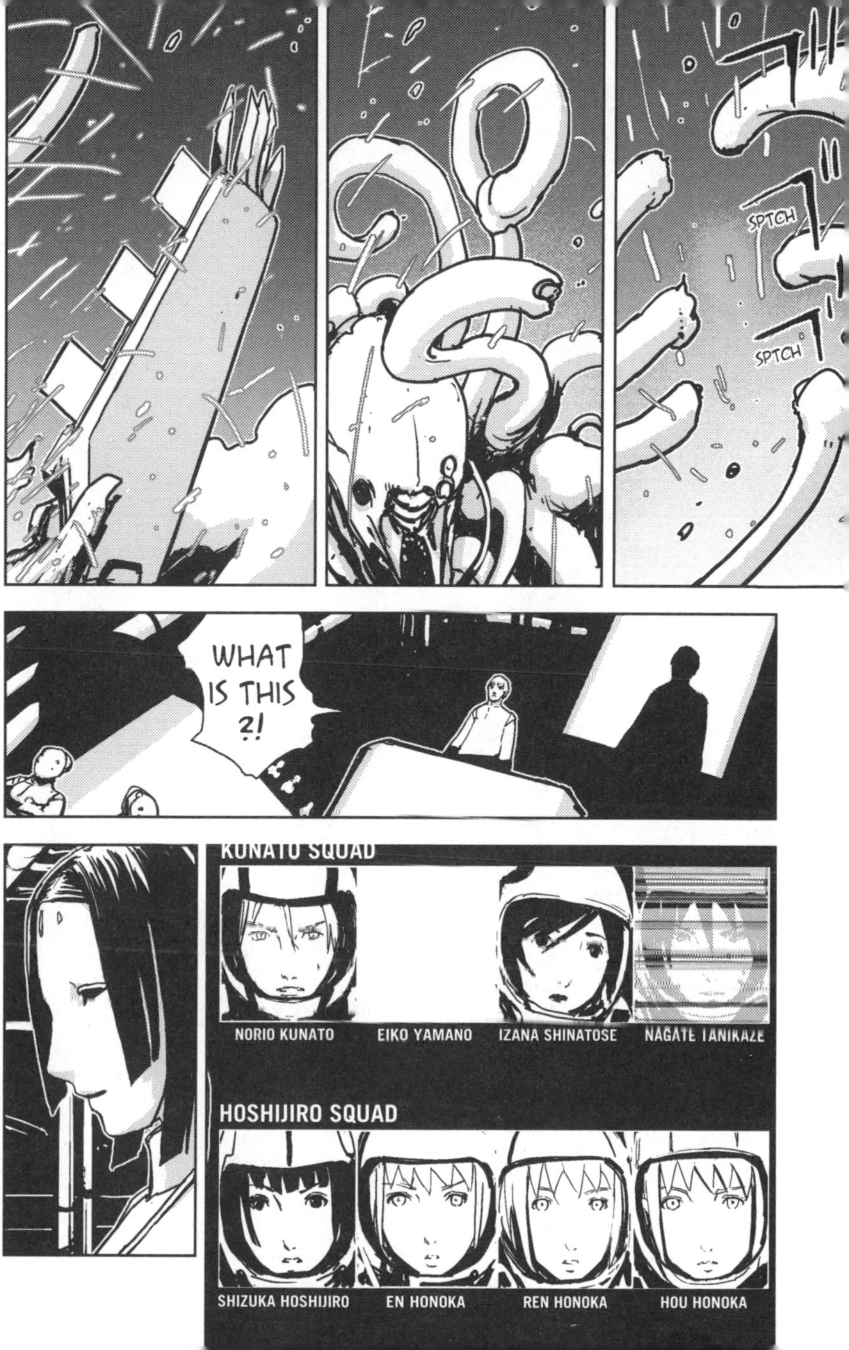
SPTCH
SPTCH
WHAT IS THIS ?!
KUNATO SQUAD
NORIO KUNATO
EIKO YAMANO
IZANA SHINATOSE
NAGATE TANIKAZE
HOSHIJIRO SQUAD
SHIZUKA HOSHIJIRO
EN HONOKA
REN HONOKA
HOU HONOKA

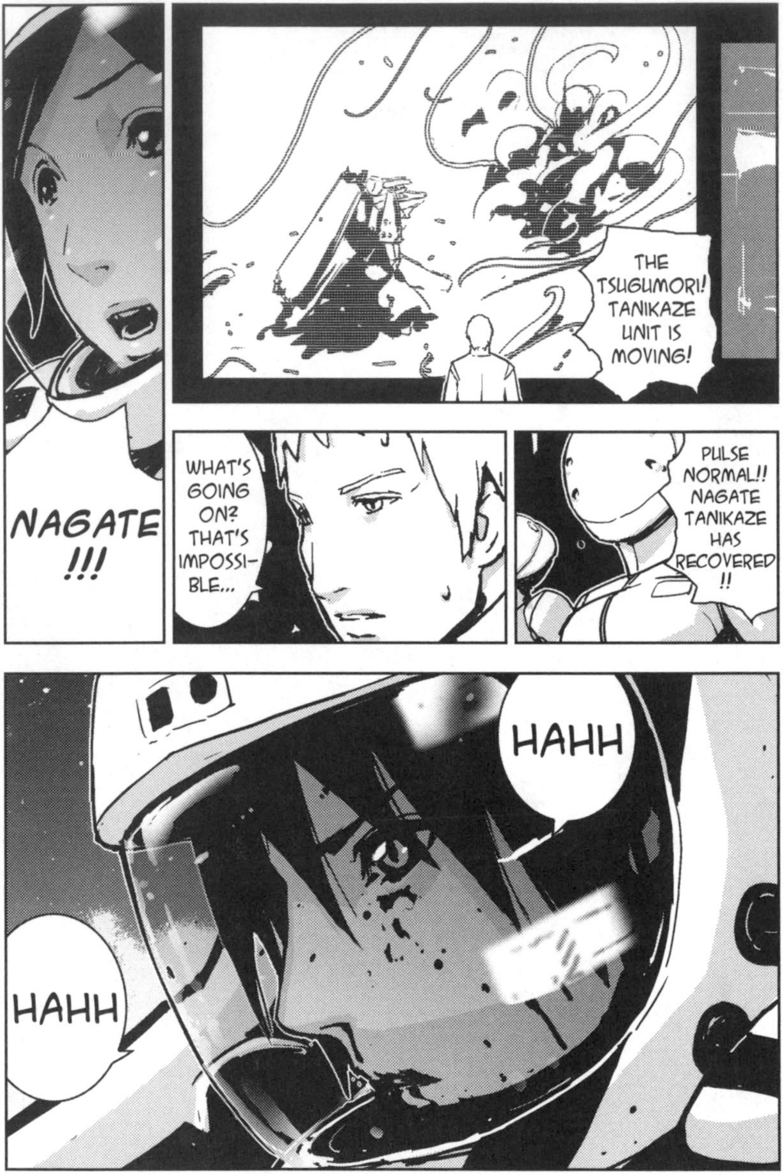
THE TSUGUMORI! TANIKAZE UNIT IS MOVING!
PULSE NORMAL!! NAGATE TANIKAZE HAS RECOVERED!!
WHAT'S GOING ON? THAT'S IMPOSSI-BLE...
NAGATE!!!
HAHH
HAHH

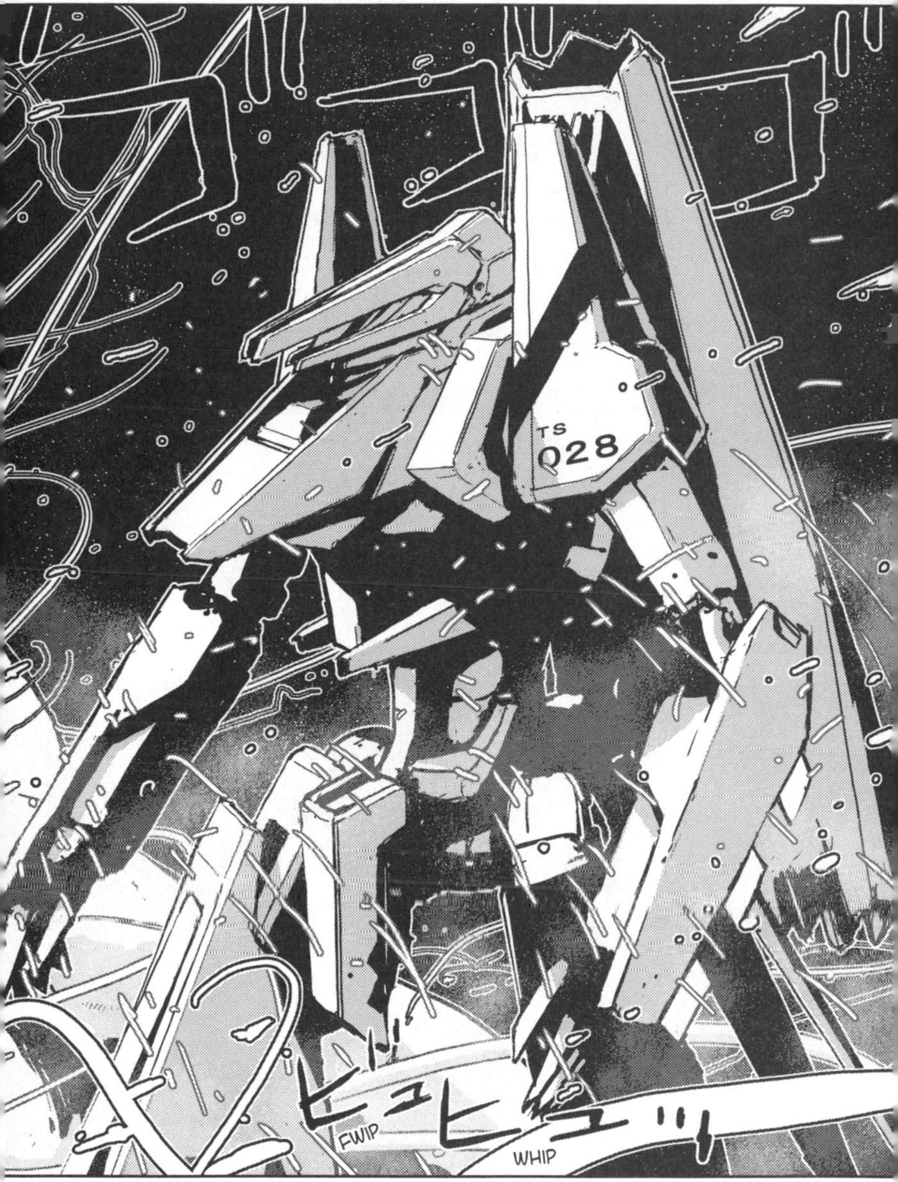
TS
028
ヒュ
FWIP
ヒュッ
WHIP

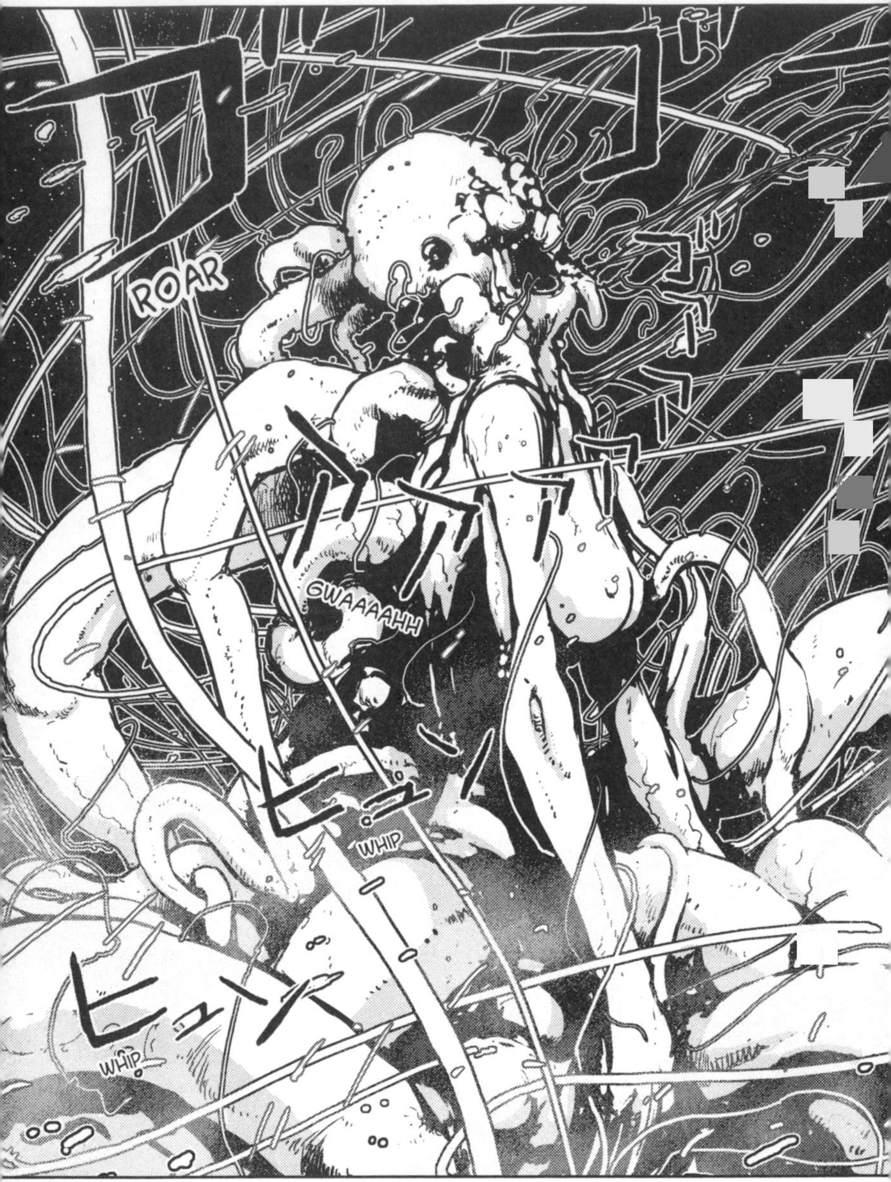
ROAR
GWAAAAHH
WHIP
WHIP

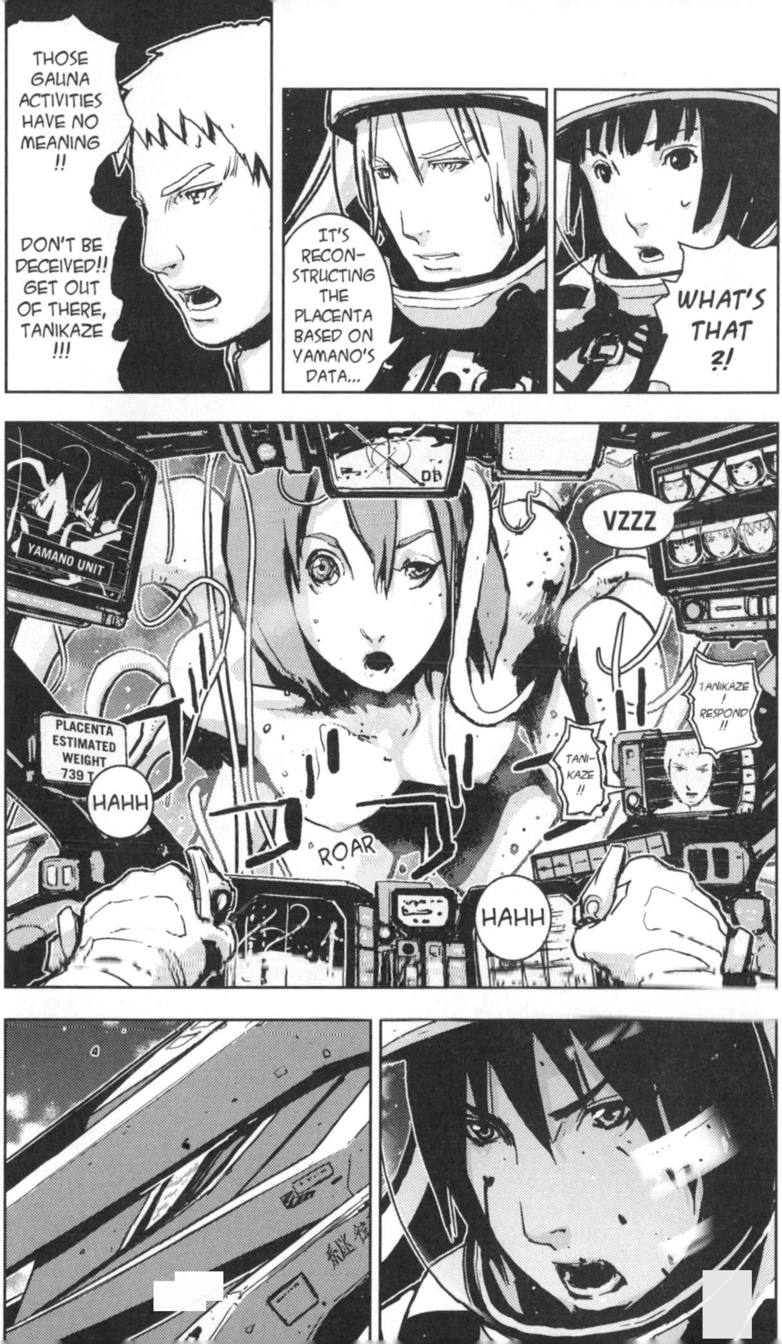
WHAT'S THAT ?!
IT'S RECON-STRUCTING THE PLACENTA BASED ON YAMANO'S DATA...
THOSE GAUNA ACTIVITIES HAVE NO MEANING !!
DON'T BE DECEIVED!! GET OUT OF THERE, TANIKAZE !!!
VZZZ
TANIKAZE ! RESPOND !!
TANI-KAZE !!
YAMANO UNIT
PLACENTA ESTIMATED WEIGHT 739 T
HAHH
HAHH
ROAR

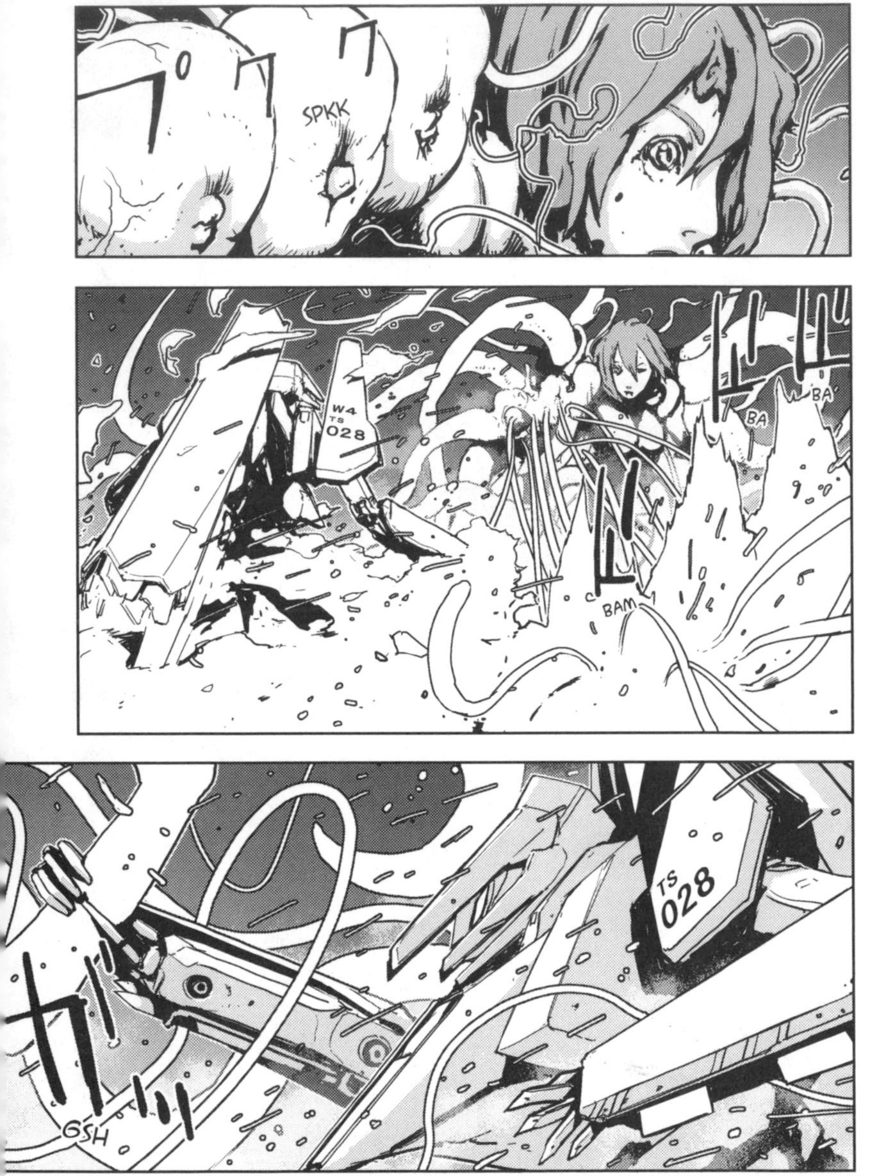
SPKK
BA
BA
BAM
GSH

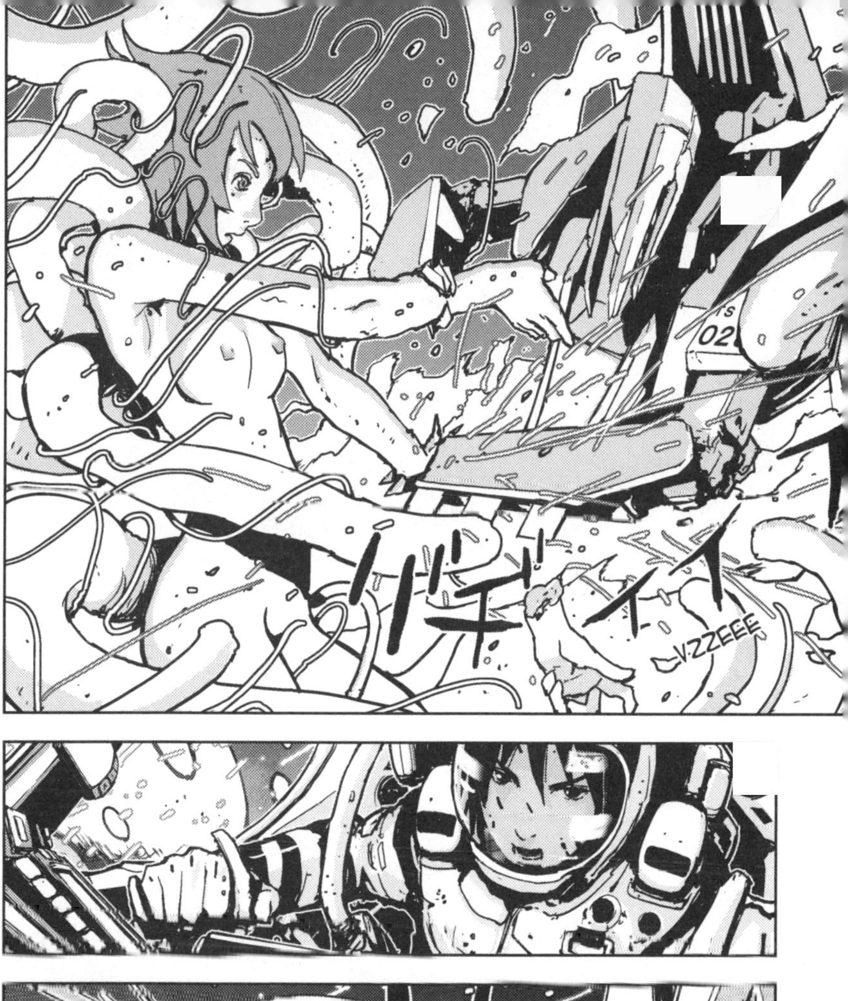
VZZEEE
BOOM
BOOM

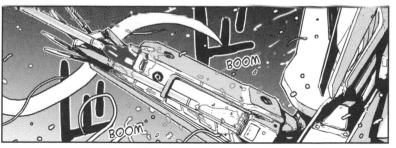

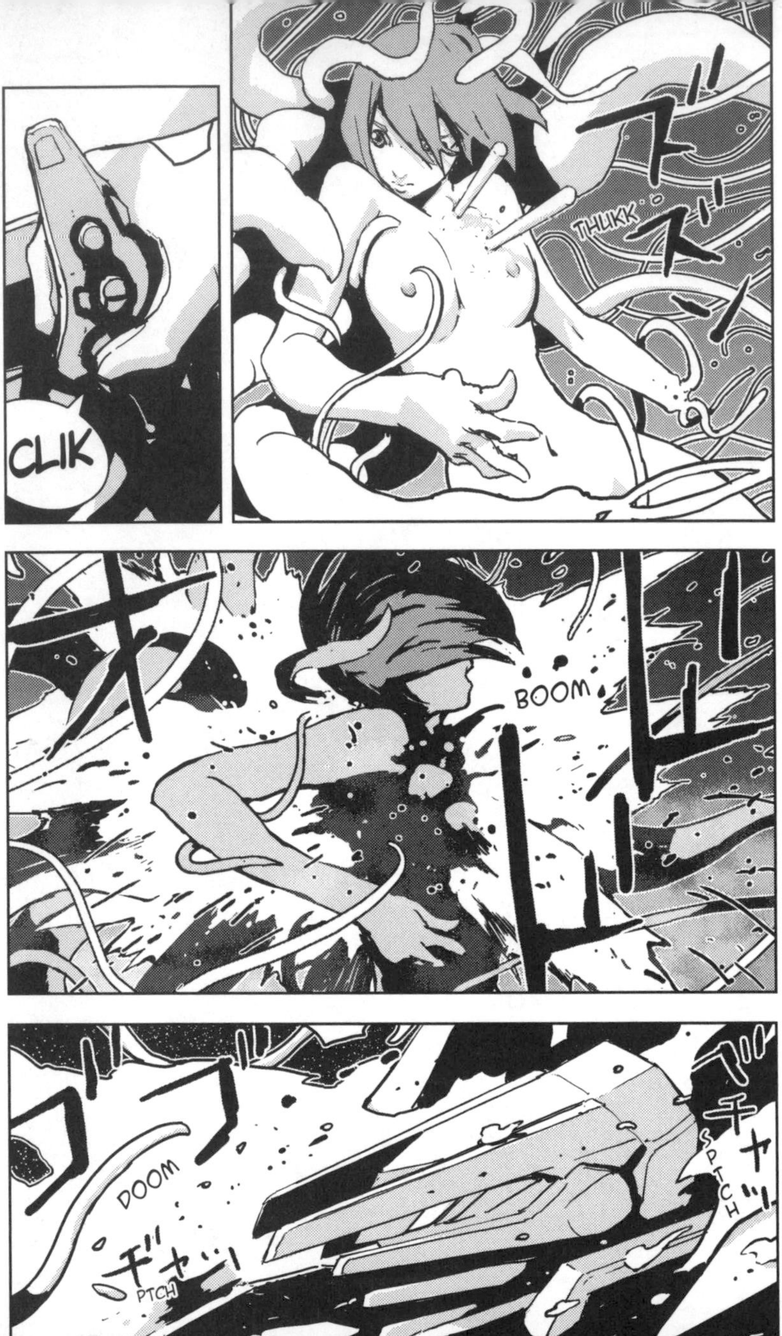
THUKK
CLIK
BOOM
SPTCH
DOOM
PTCH

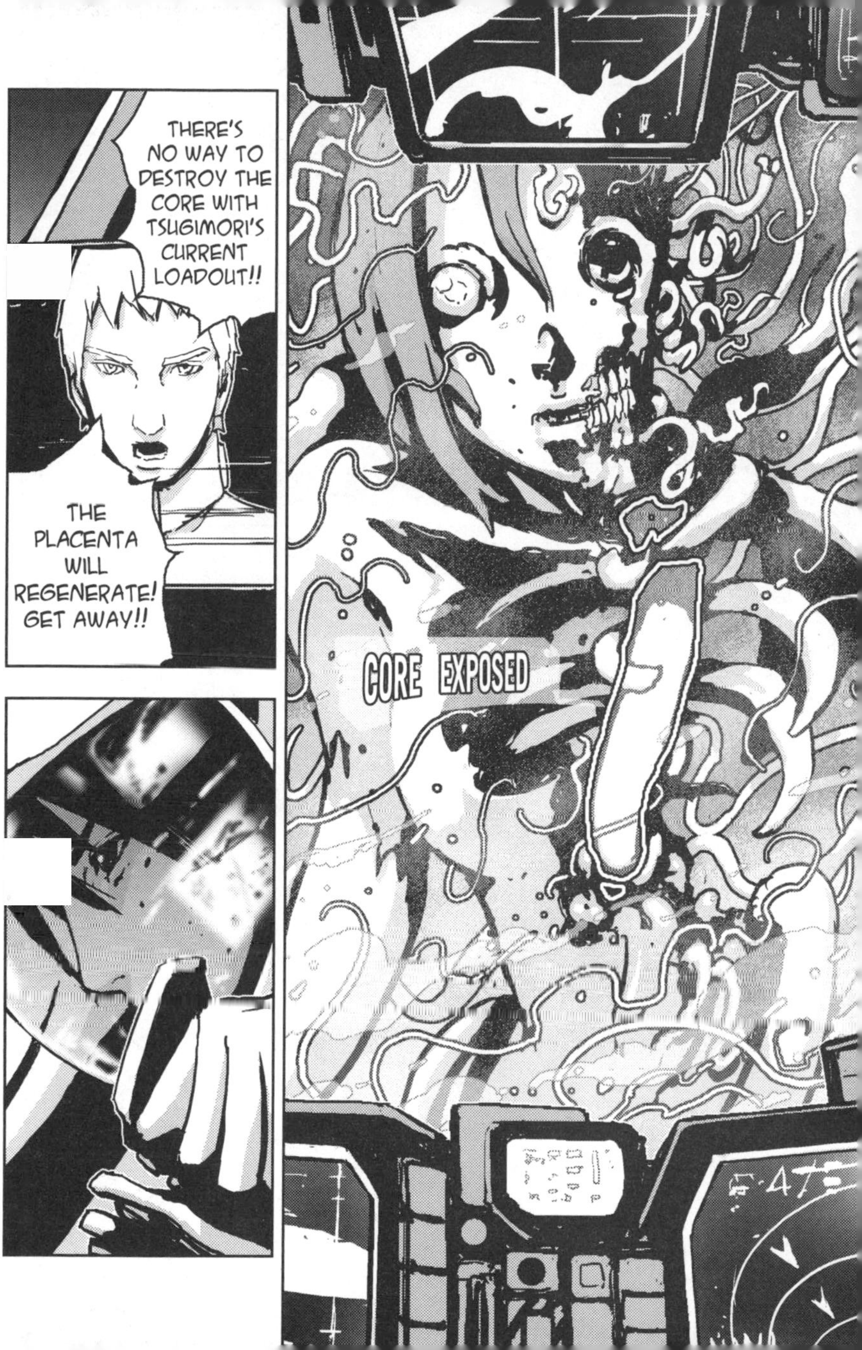
CORE EXPOSED
THERE'S NO WAY TO DESTROY THE CORE WITH TSUGIMORI'S CURRENT LOADOUT!!
THE PLACENTA WILL REGENERATE! GET AWAY!!

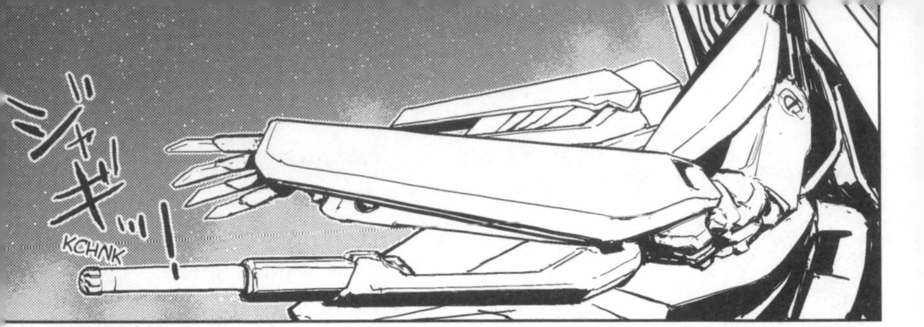
ジャギッ!!
KCHNK
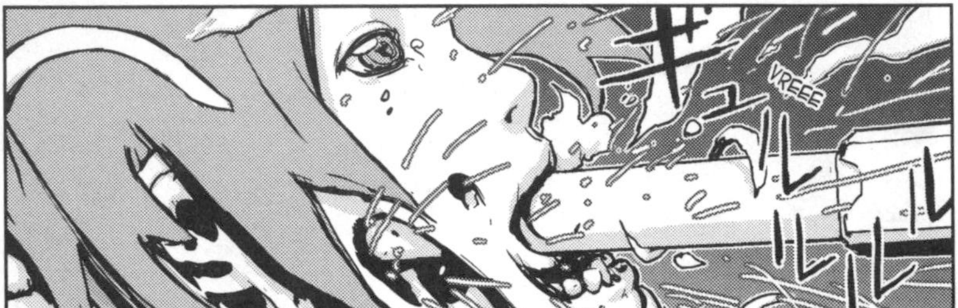
ギュルルルル
VREEE
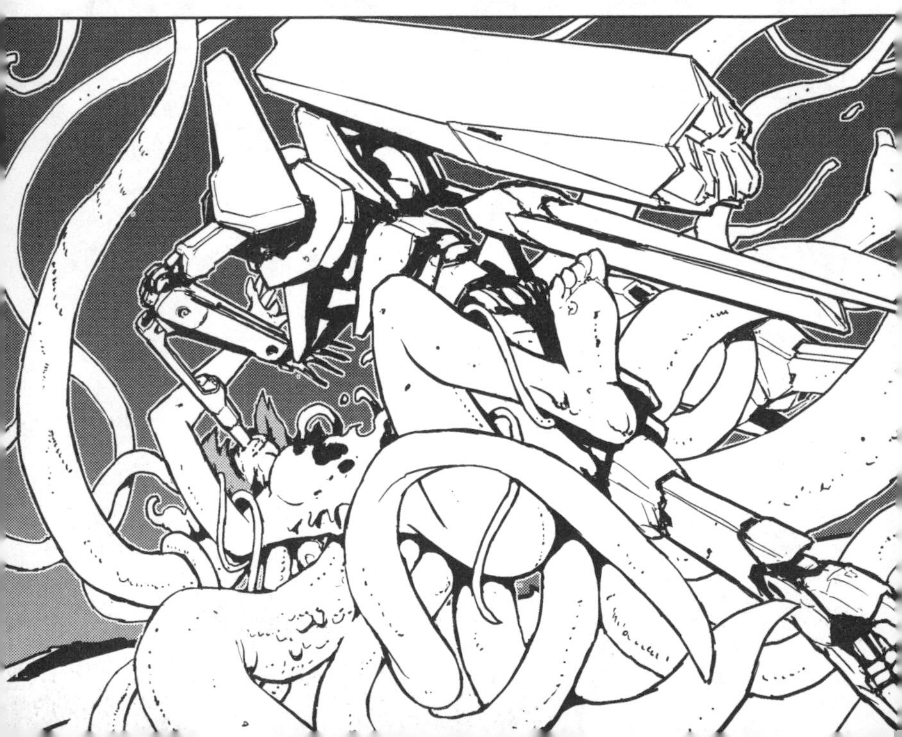

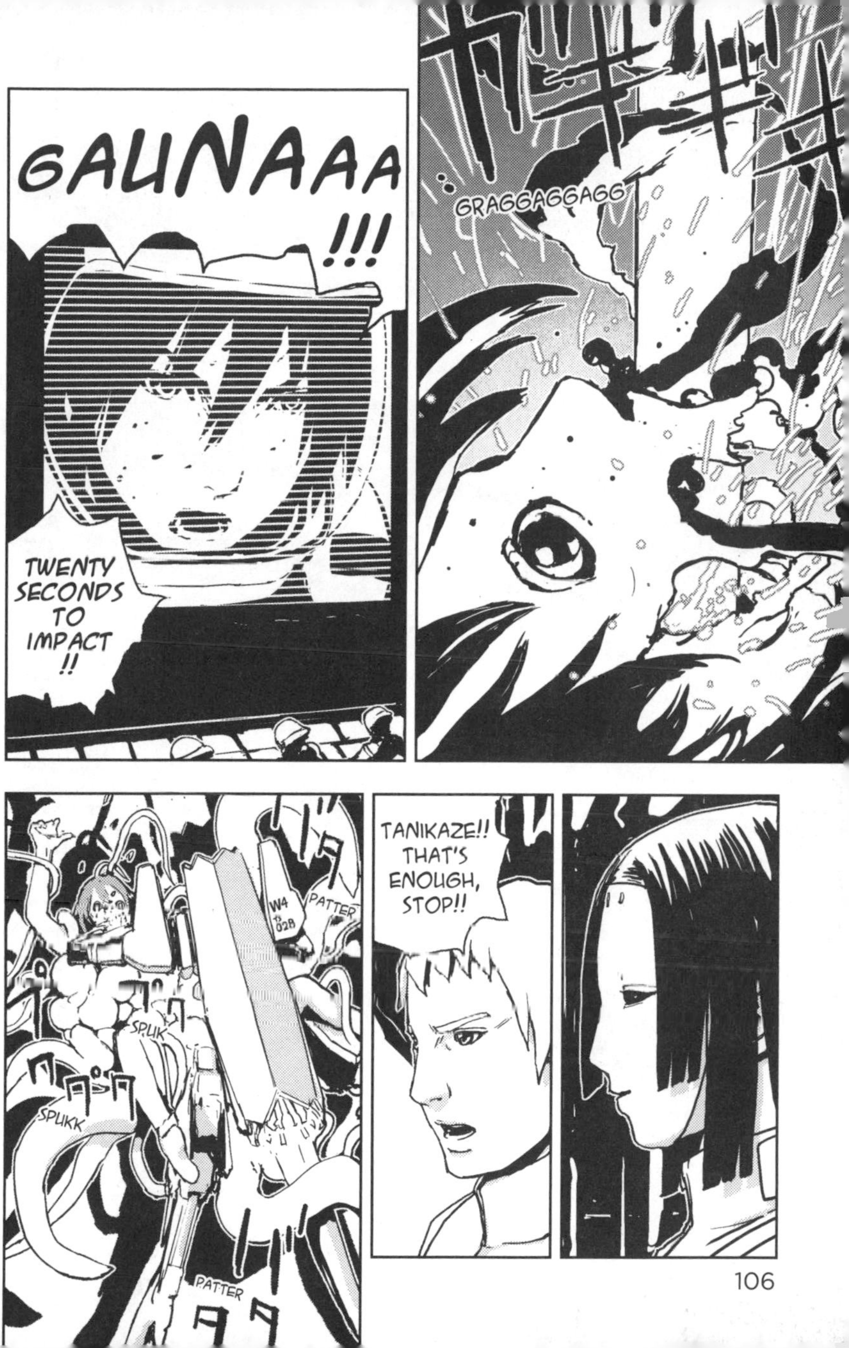
GRAGGAGGAGG
GAUNAAA!!!
TWENTY SECONDS TO IMPACT!!
TANIKAZE!! THAT'S ENOUGH, STOP!!
PATTER
W4
028
SPUK
SPUKK
PATTER

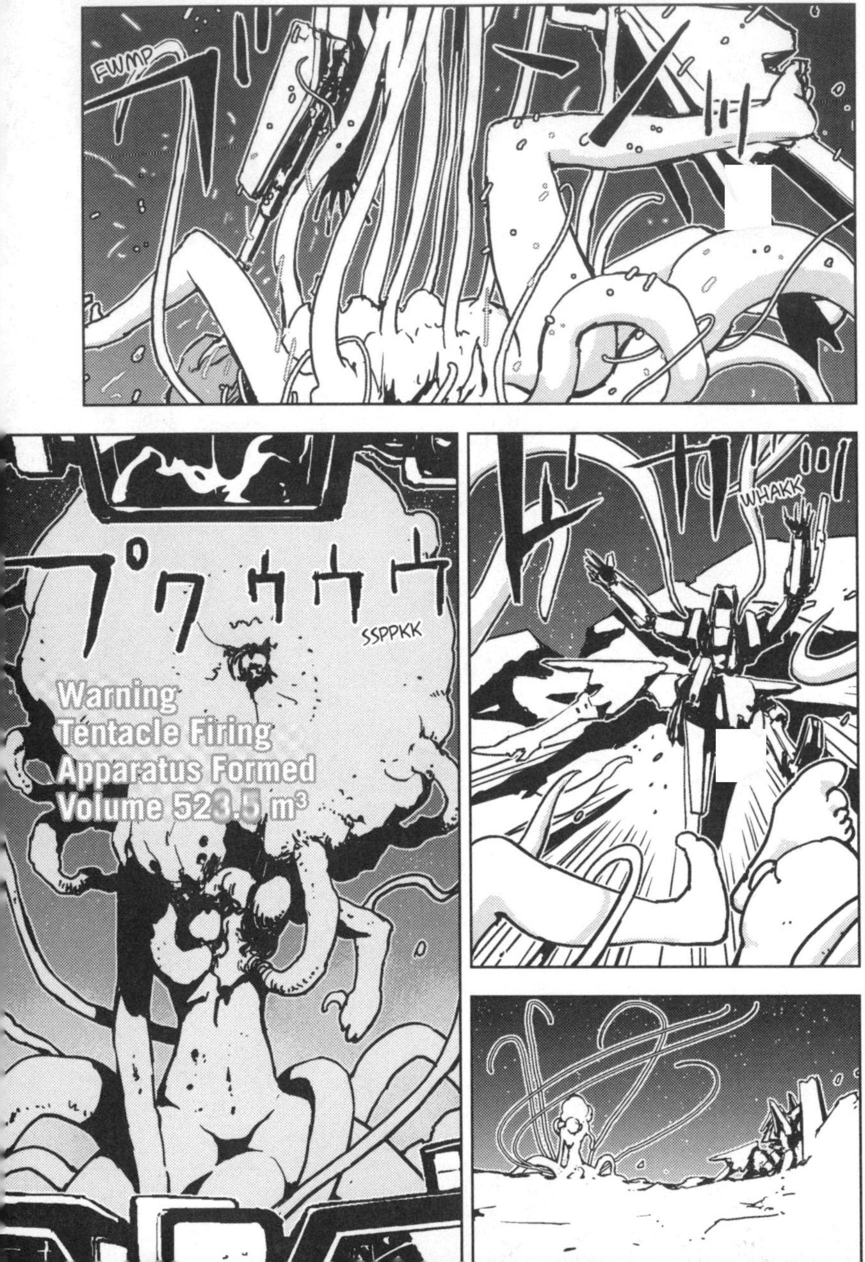
FWMP
WHAKK
プクウウウ
SSPPKK
Warning
Tentacle Firing
Apparatus Formed
Volume 523.5 m³

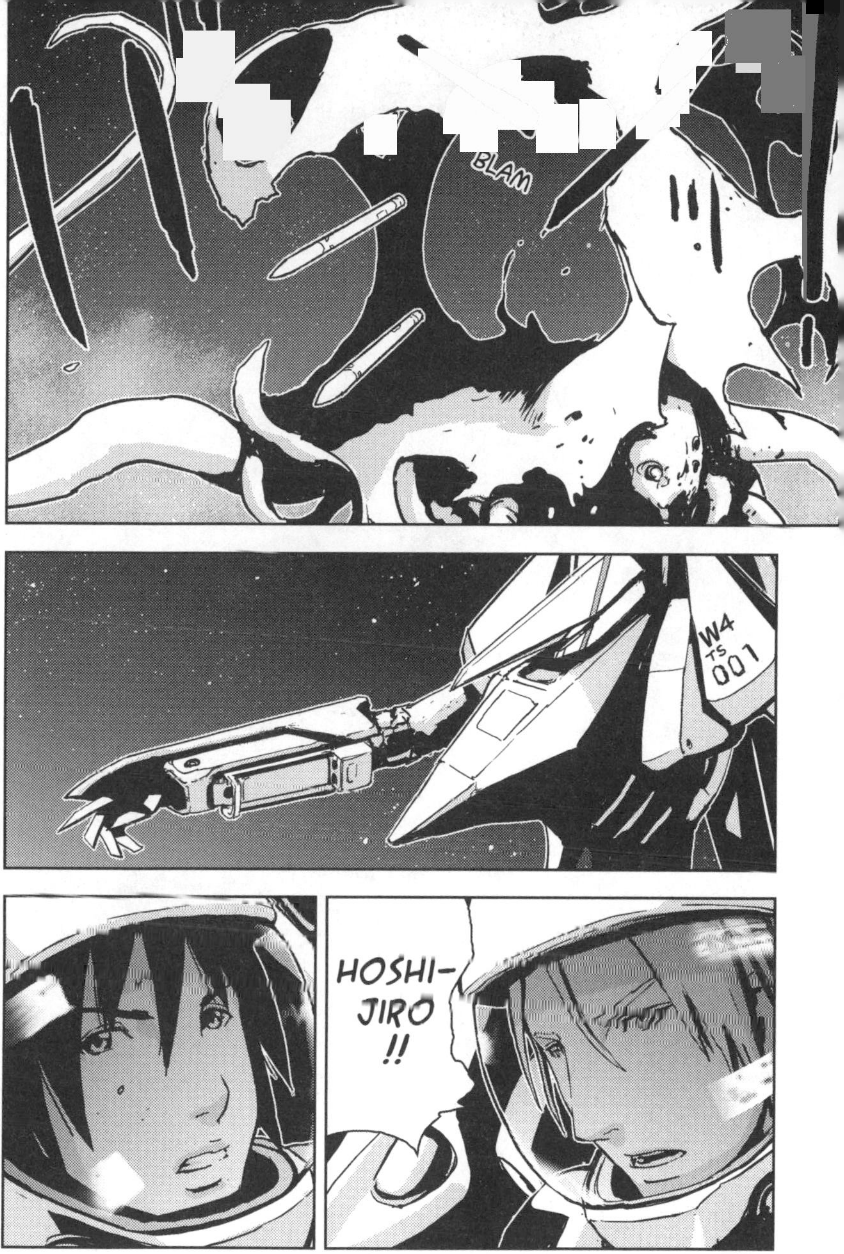
BLAM
W4
TS
001
HOSHI-
JIRO
!!

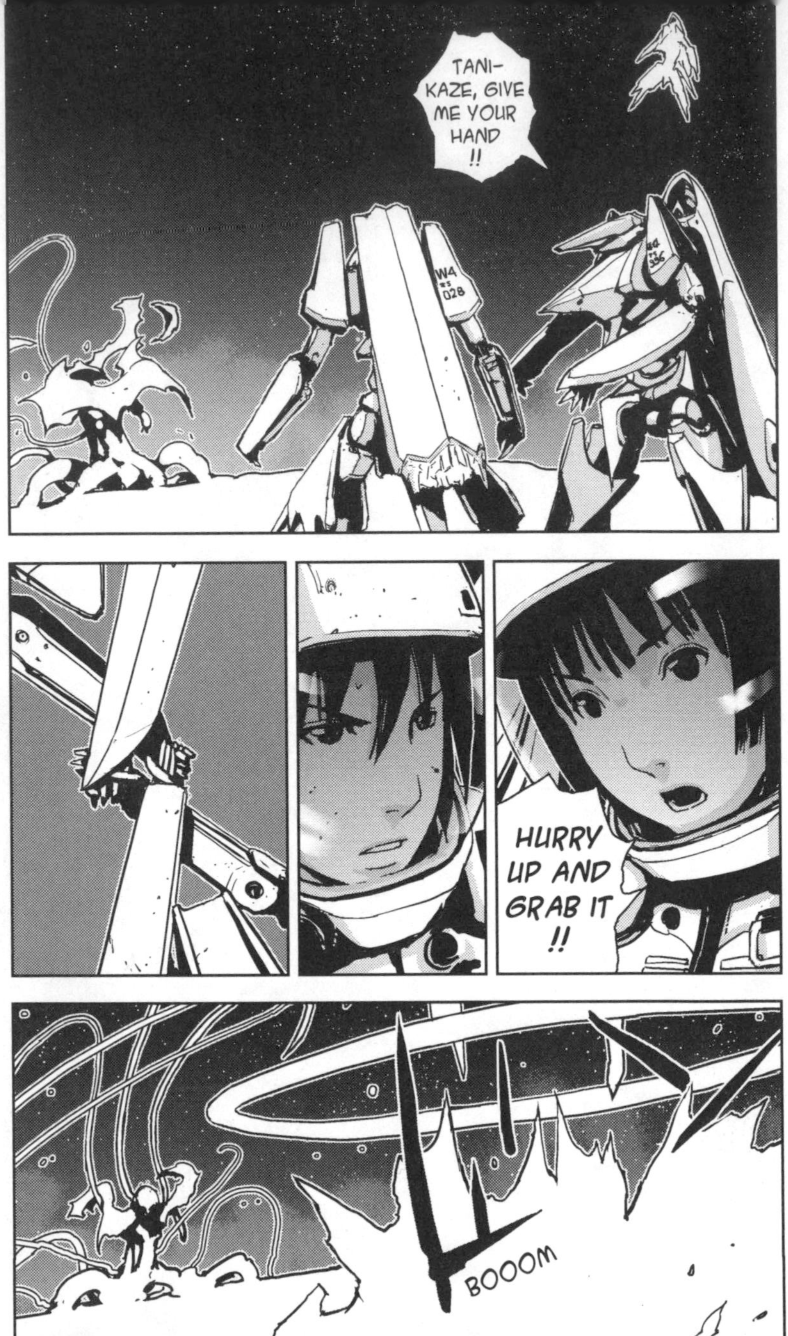
TANI-KAZE, GIVE ME YOUR HAND !!
HURRY UP AND GRAB IT !!
BOOOM

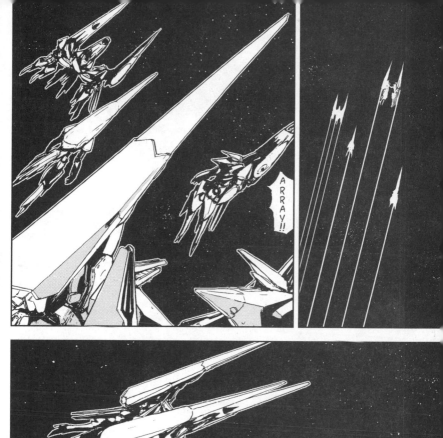

ARRAY!!

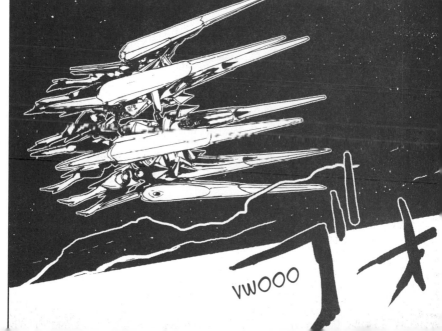

VWOOO

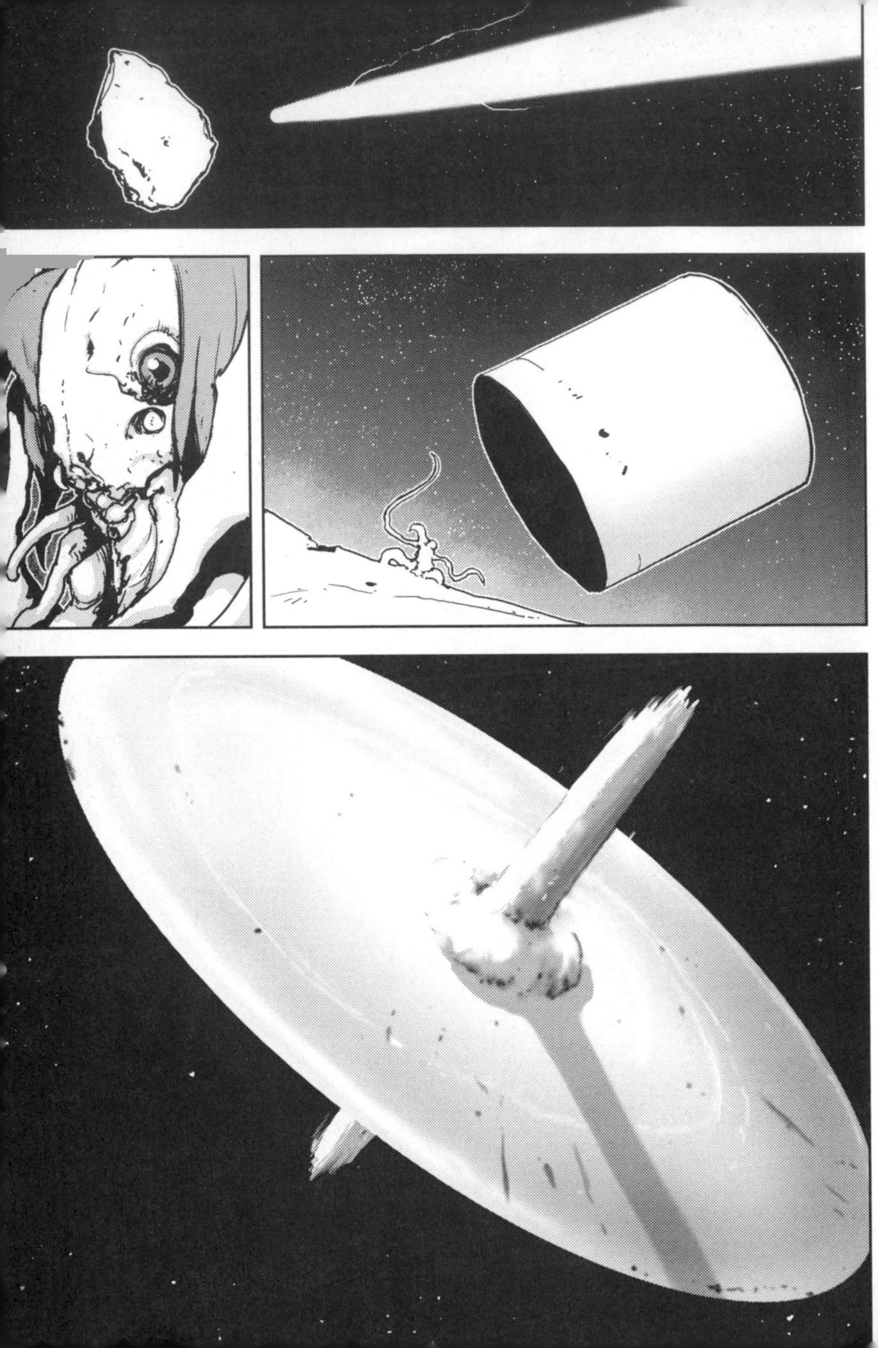

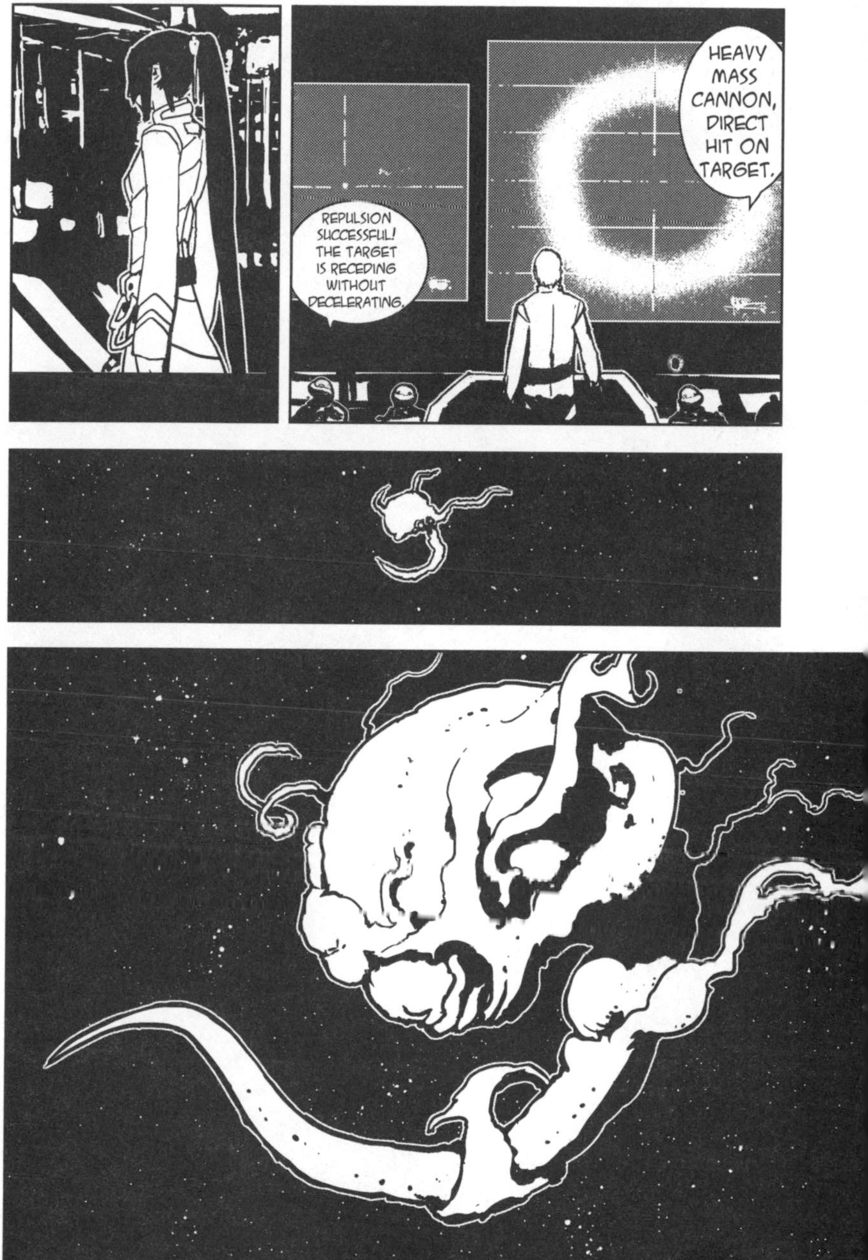
HEAVY MASS CANNON, DIRECT HIT ON TARGET.
REPULSION SUCCESSFUL! THE TARGET IS RECEDING WITHOUT DECELERATING.

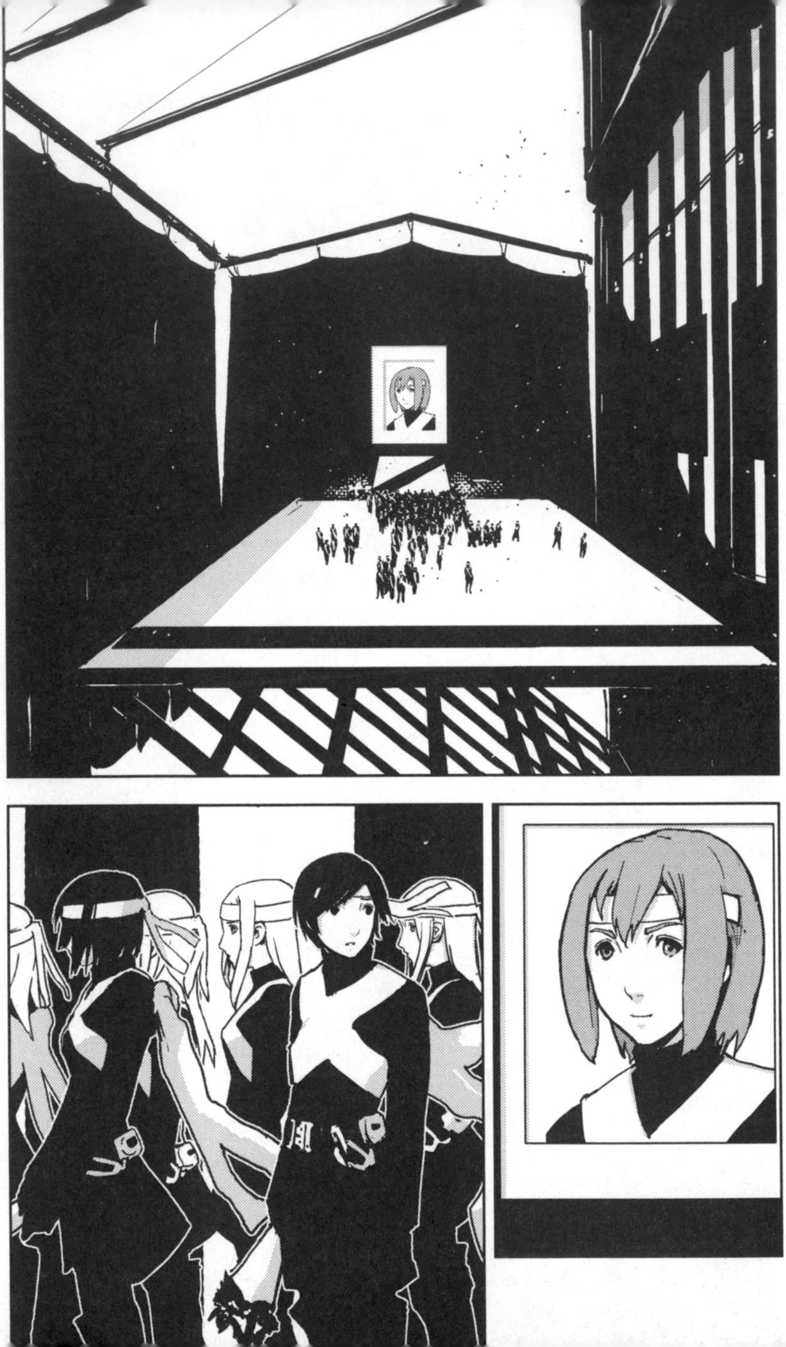

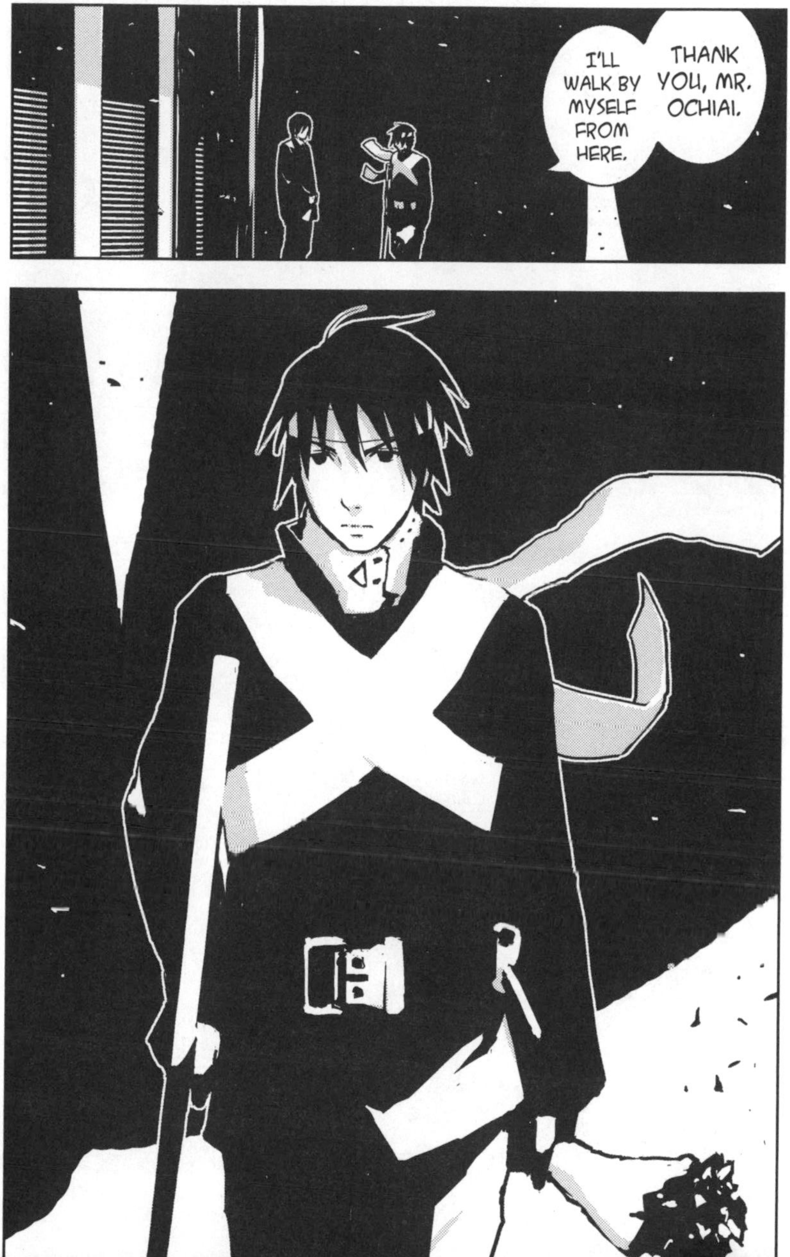
THANK YOU, MR. OCHIAI.
I'LL WALK BY MYSELF FROM HERE.

Chapter 3: END

Chapter 4: Norio Kunato's Fury

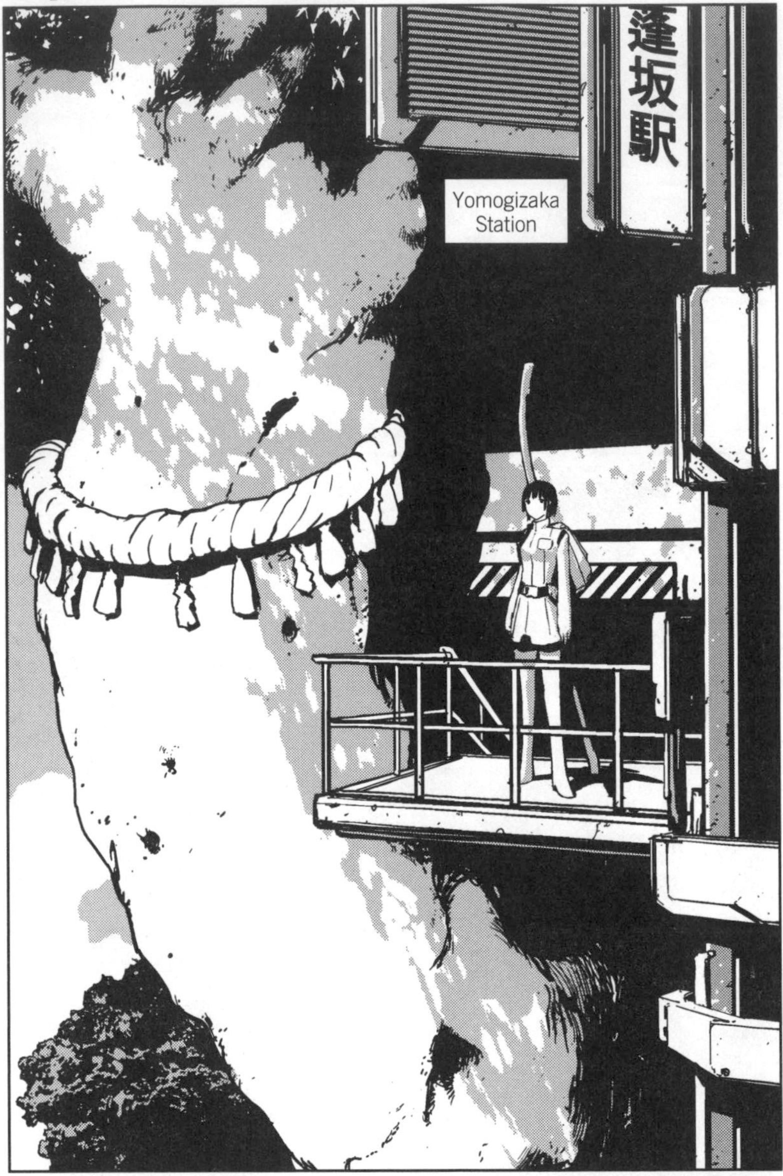
蓬坂駅
Yomogizaka Station

One Hundred Sights of Sidonia Part Three:
High Speed Persons and Freight Elevator Track

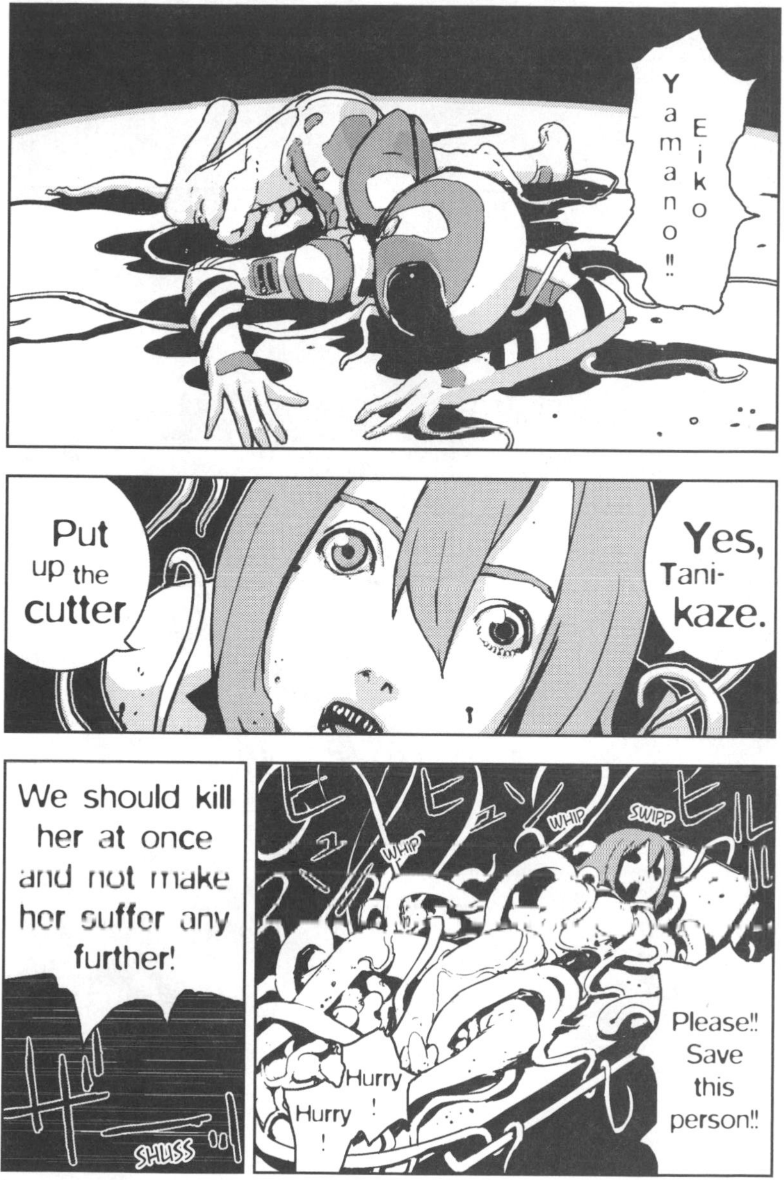
Eiko Yamano!!
Yes, Tani-kaze.
Put up the cutter
SWIPP
WHIP
WHIP
Please!! Save this person!!
Hurry!
Hurry!
We should kill her at once and not make her suffer any further!
SHUSS

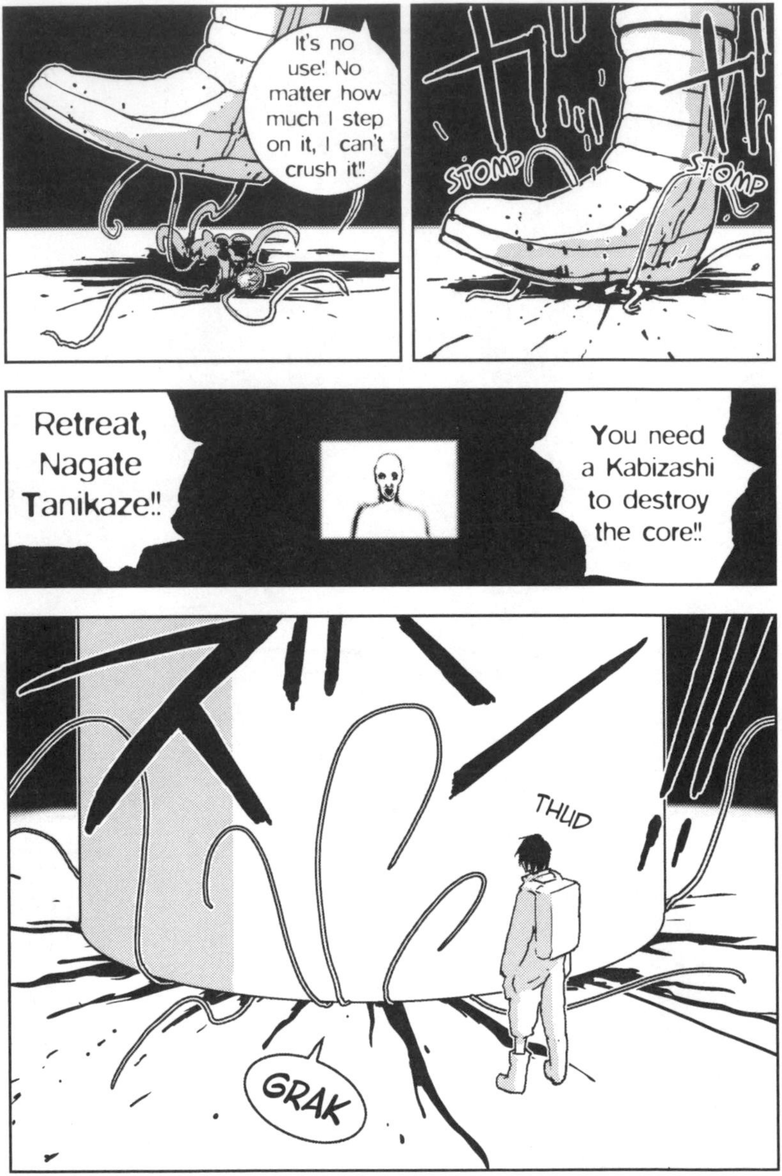
STOMP
STOMP
It's no use! No matter how much I step on it, I can't crush it!!
You need a Kabizashi to destroy the core!!
Retreat, Nagate Tanikaze!!
THUD
GRAK

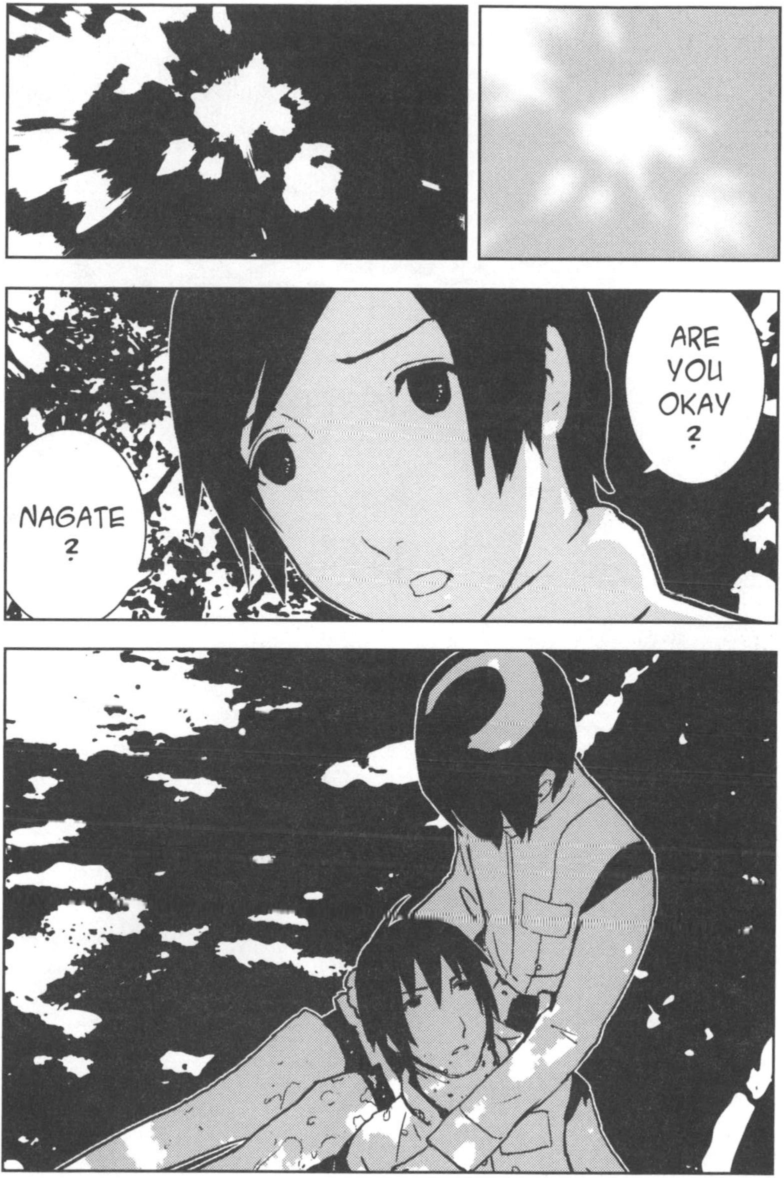
ARE YOU OKAY?
NAGATE?

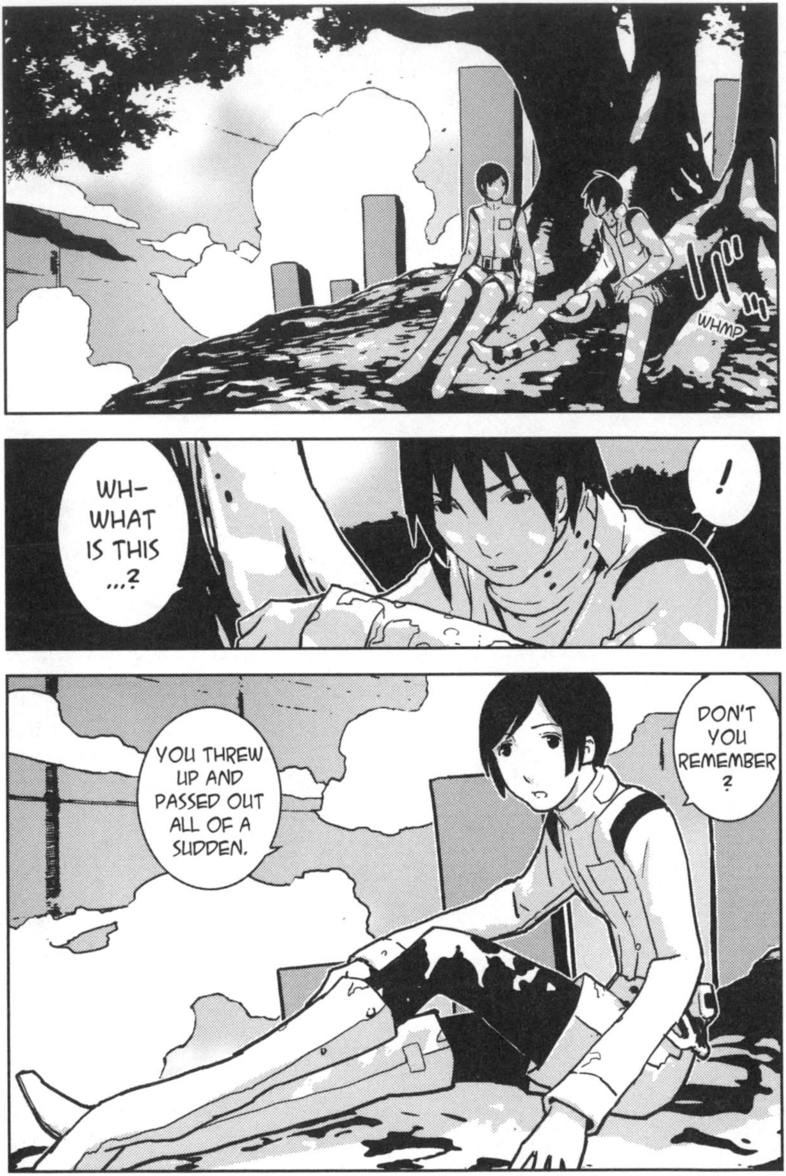
WHMP
!
WH- WHAT IS THIS ...?
DON'T YOU REMEMBER?
YOU THREW UP AND PASSED OUT ALL OF A SUDDEN.

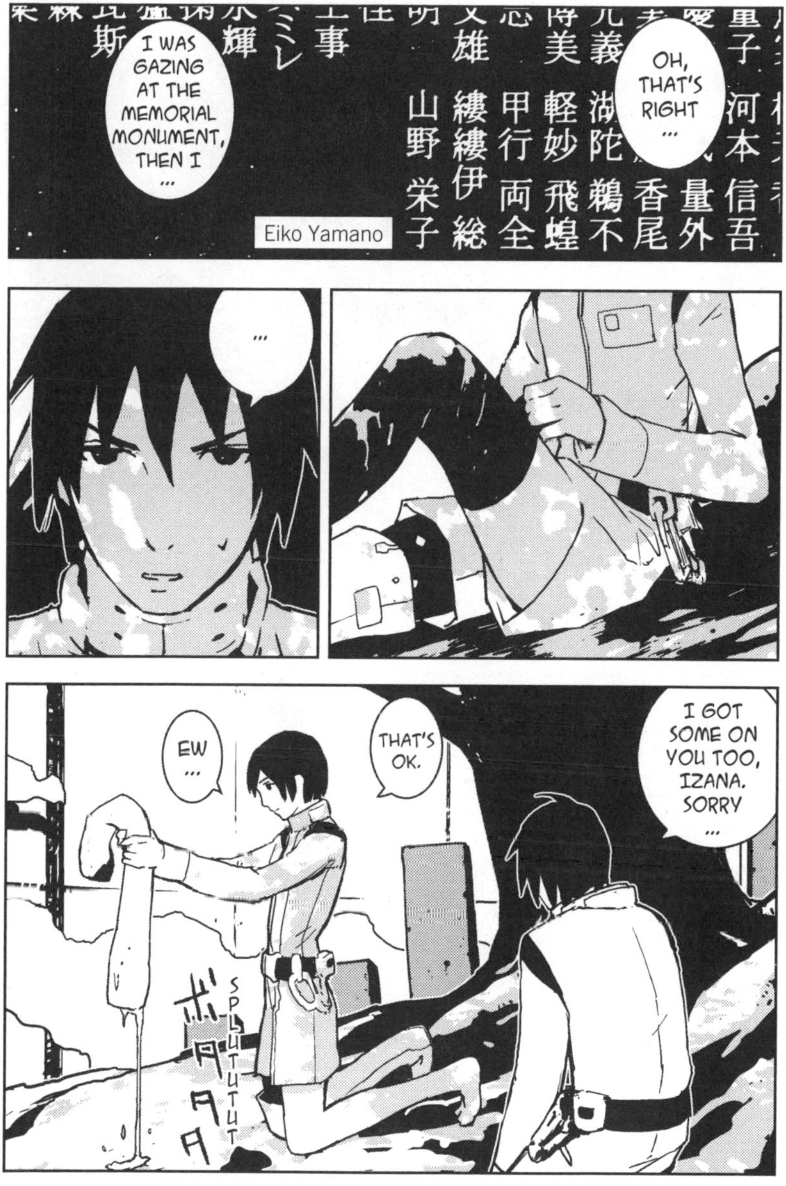

OH, THAT'S RIGHT ...
河本信吾
軽妙飛蝗
甲行両全
縷縷伊総
山野栄子
I WAS GAZING AT THE MEMORIAL MONUMENT, THEN I ...
Eiko Yamano
...
I GOT SOME ON YOU TOO, IZANA. SORRY ...
THAT'S OK.
EW ...
SPLUTUTUT

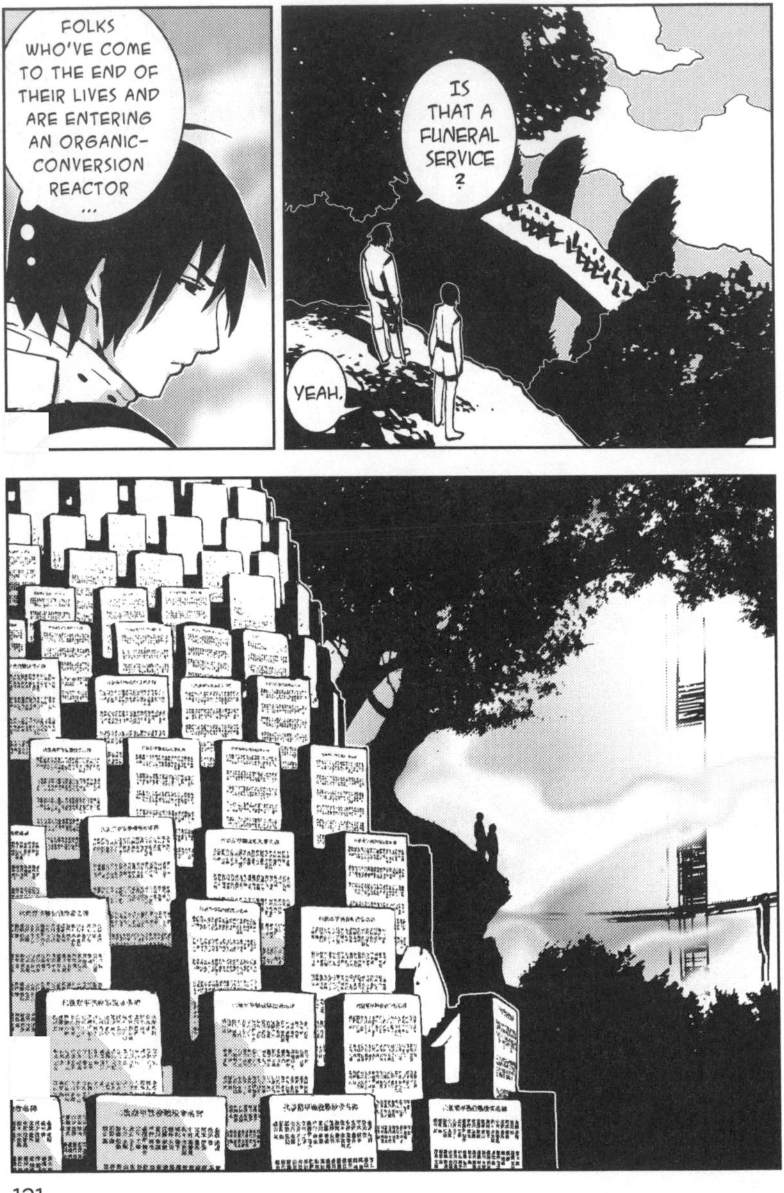
IS THAT A FUNERAL SERVICE?
YEAH.
FOLKS WHO'VE COME TO THE END OF THEIR LIVES AND ARE ENTERING AN ORGANIC-CONVERSION REACTOR ...

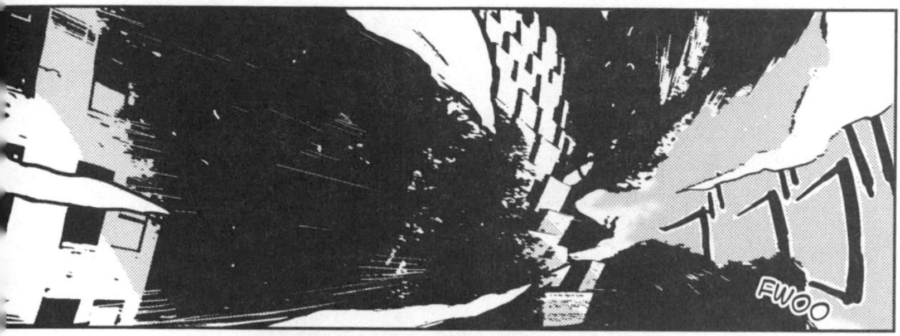
FWOO
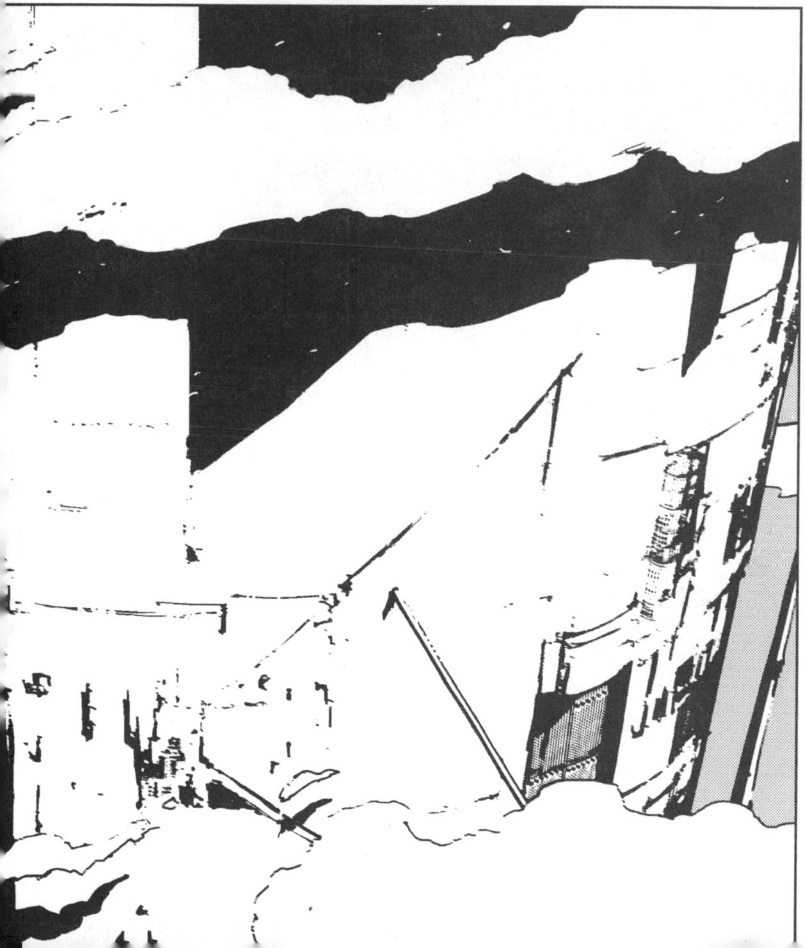

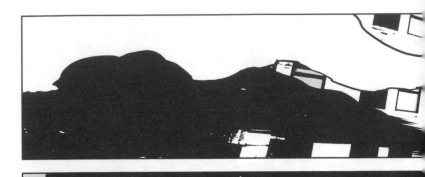

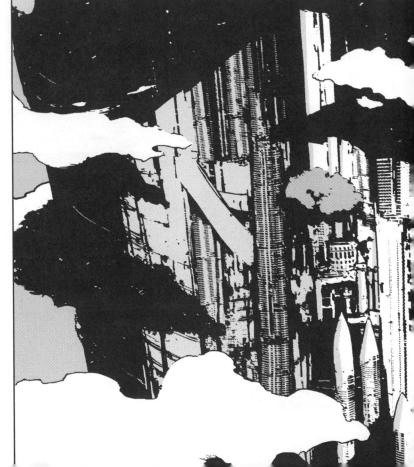

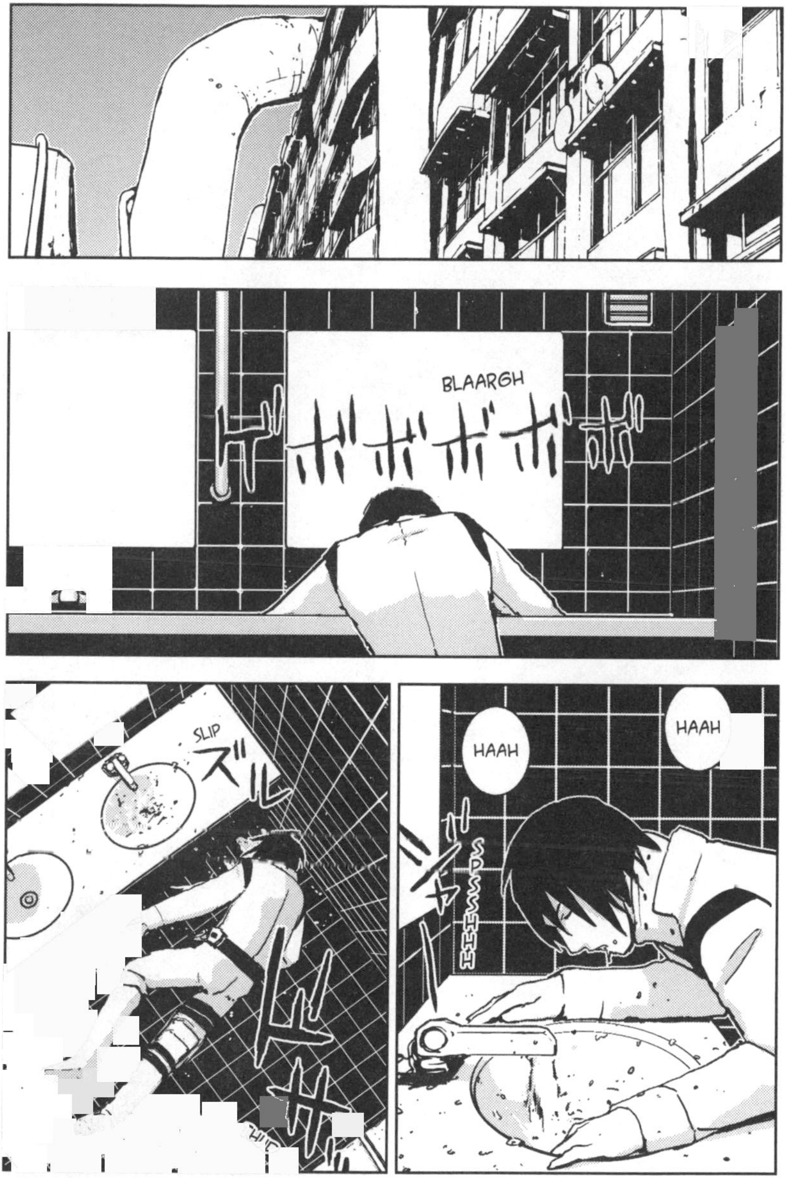
BLAARGH
ゲボボボボボ
HAAH
HAAH
SLIP
ズル

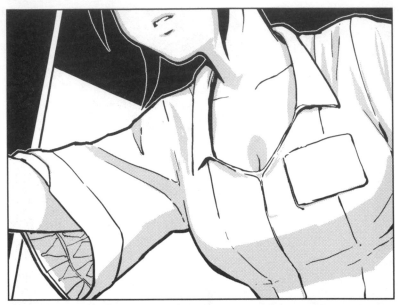

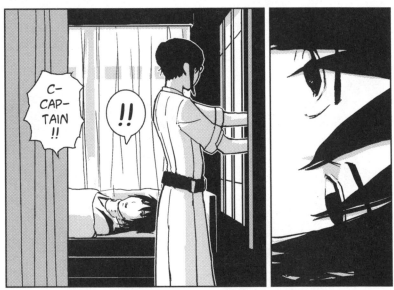

C-CAP-TAIN!!

!!

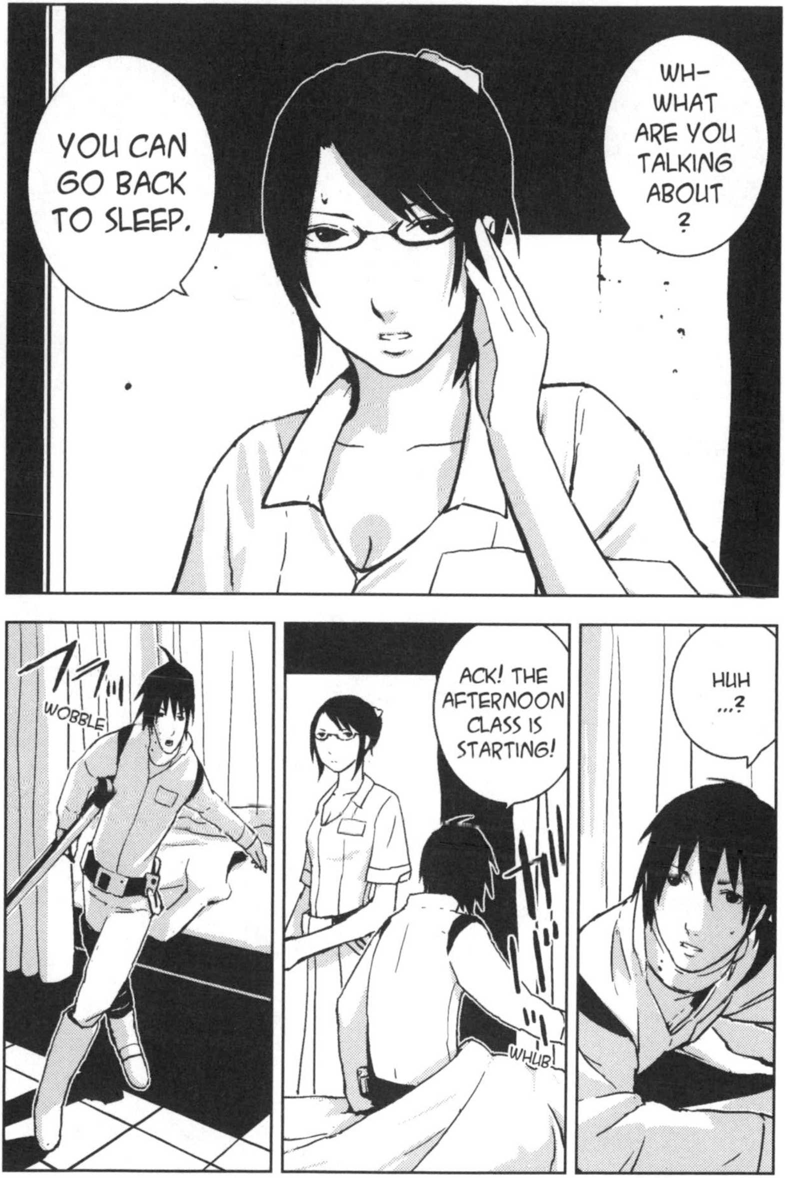
WH- WHAT ARE YOU TALKING ABOUT?
YOU CAN GO BACK TO SLEEP.
HUH ...?
ACK! THE AFTERNOON CLASS IS STARTING!
WHUB!
WOBBLE

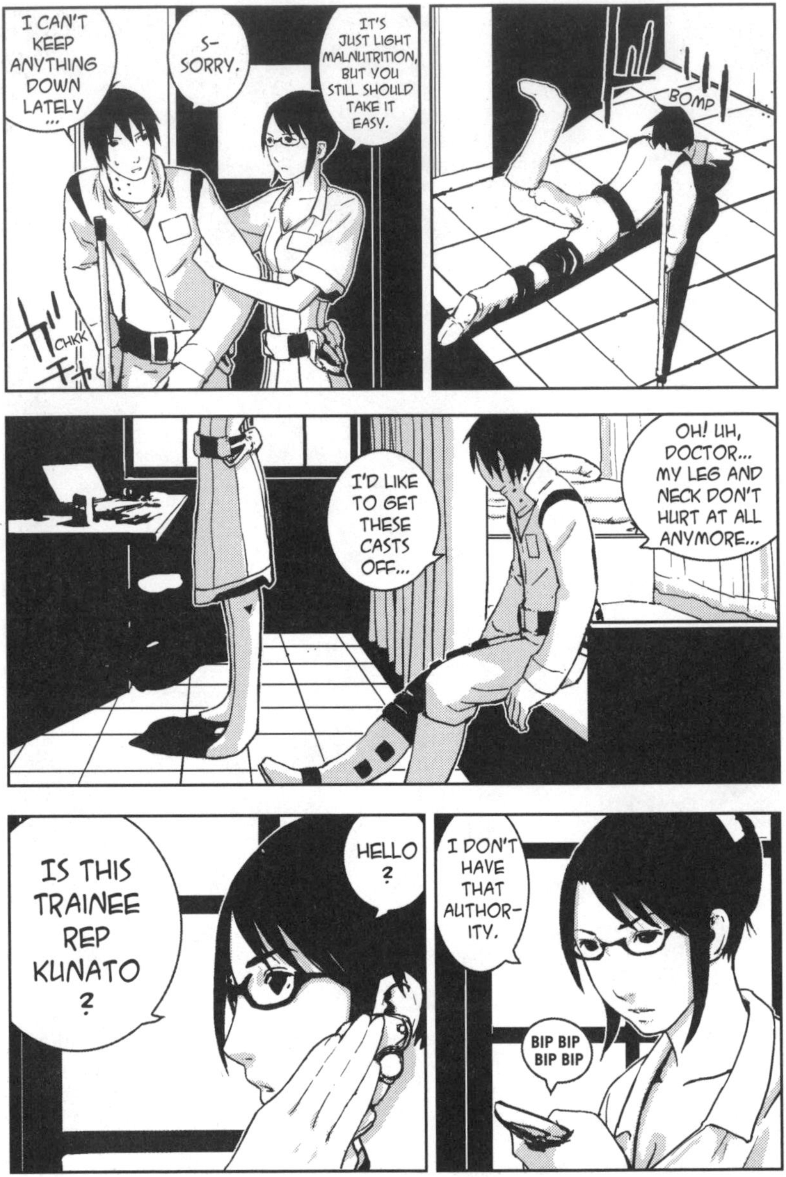
BOMP
IT'S JUST LIGHT MALNUTRITION, BUT YOU STILL SHOULD TAKE IT EASY.
S-SORRY.
I CAN'T KEEP ANYTHING DOWN LATELY ...
ガチャ CHKK
OH! UH, DOCTOR... MY LEG AND NECK DON'T HURT AT ALL ANYMORE...
I'D LIKE TO GET THESE CASTS OFF...
I DON'T HAVE THAT AUTHORITY.
BIP BIP BIP BIP
HELLO?
IS THIS TRAINEE REP KUNATO?

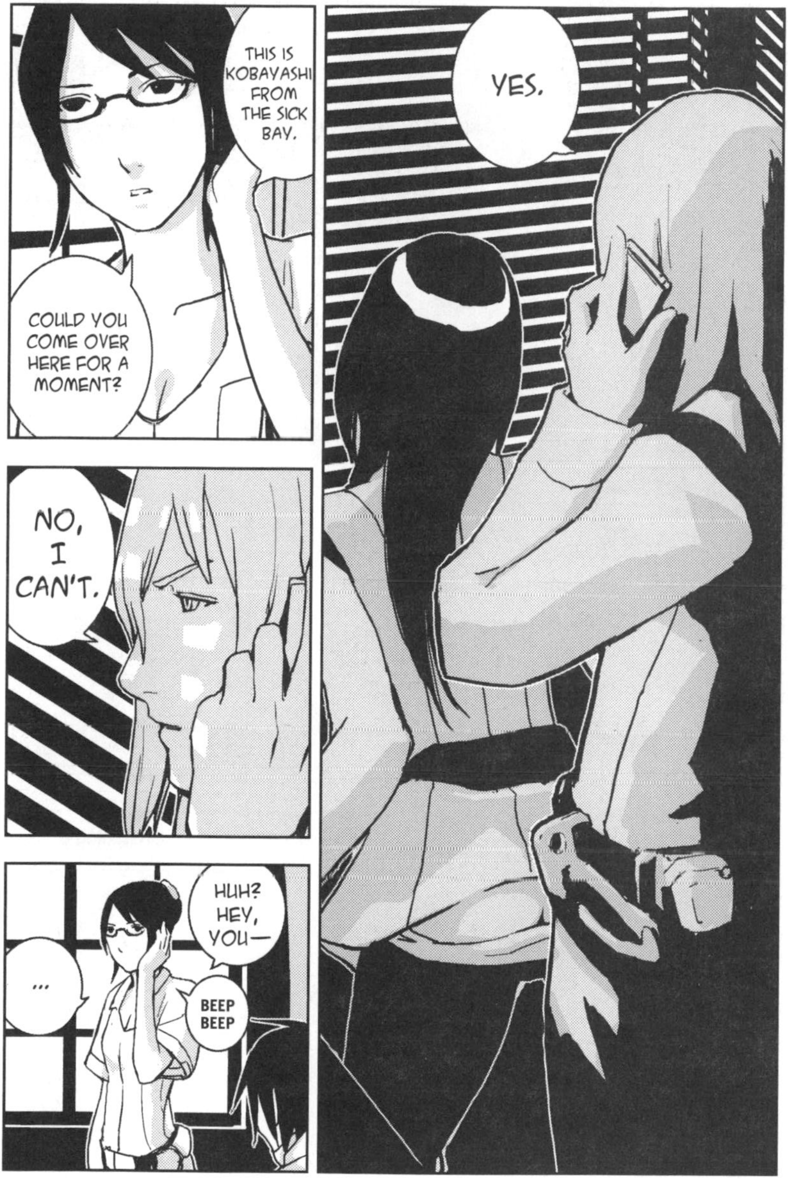
YES.
THIS IS KOBAYASHI FROM THE SICK BAY.
COULD YOU COME OVER HERE FOR A MOMENT?
NO, I CAN'T.
HUH? HEY, YOU—
BEEP BEEP
...

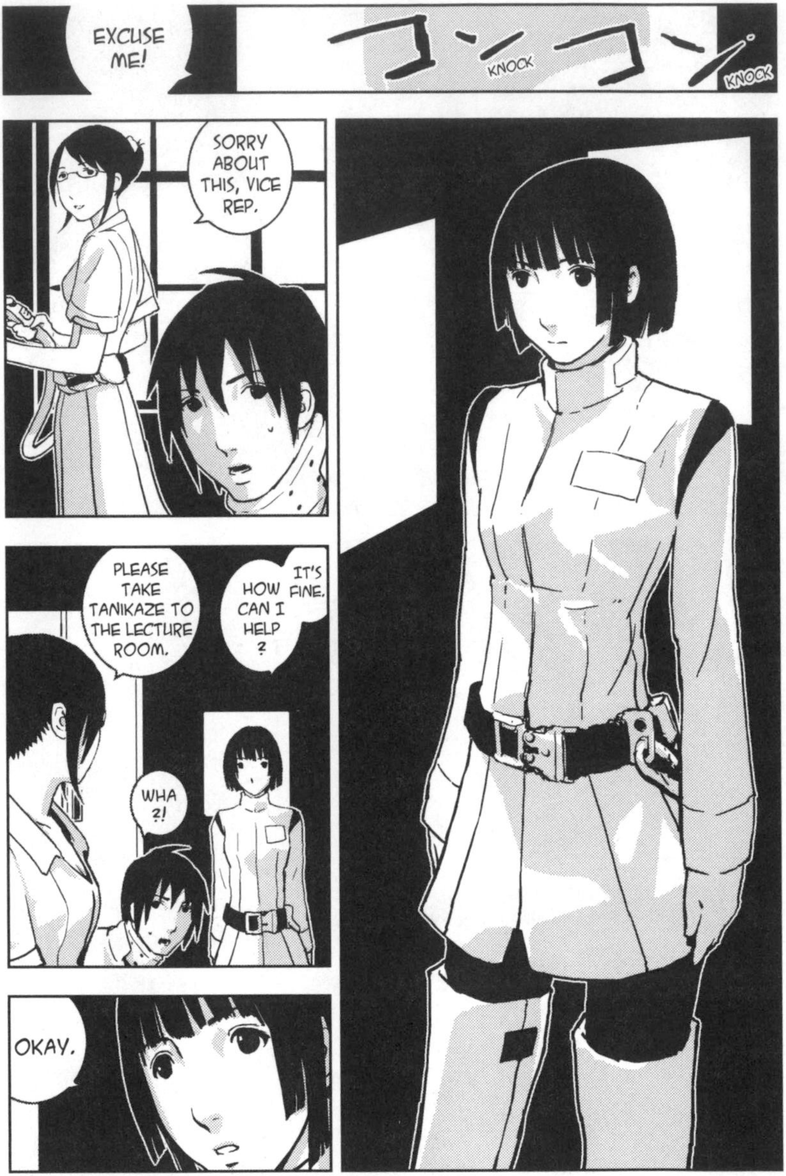
コンコン
KNOCK
KNOCK
EXCUSE ME!
SORRY ABOUT THIS, VICE REP.
IT'S FINE.
HOW CAN I HELP?
PLEASE TAKE TANIKAZE TO THE LECTURE ROOM.
WHA?!
OKAY.

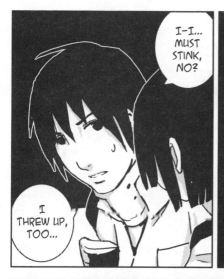

I-I... MUST STINK, NO?

I THREW UP, TOO...

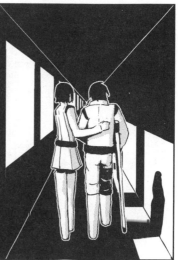

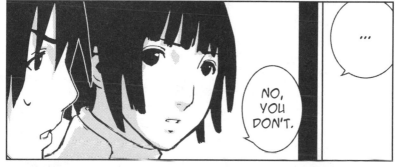

...

NO, YOU DON'T.

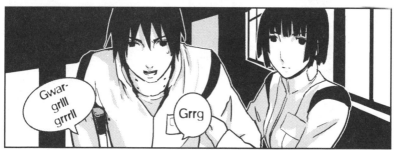

Gwar-grlll grrrll

Grrg

130

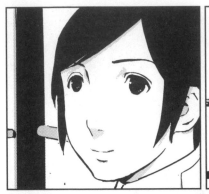

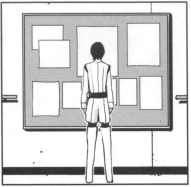

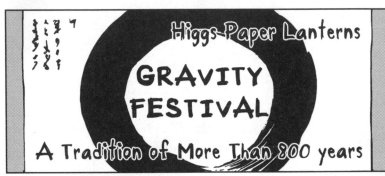

Higgs Paper Lanterns

GRAVITY FESTIVAL

A Tradition of More Than 800 years

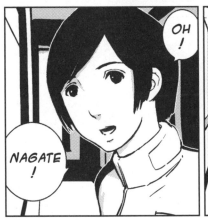

OH
!

NAGATE
!

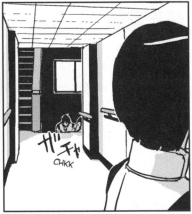

ガチャ
CHKK

131

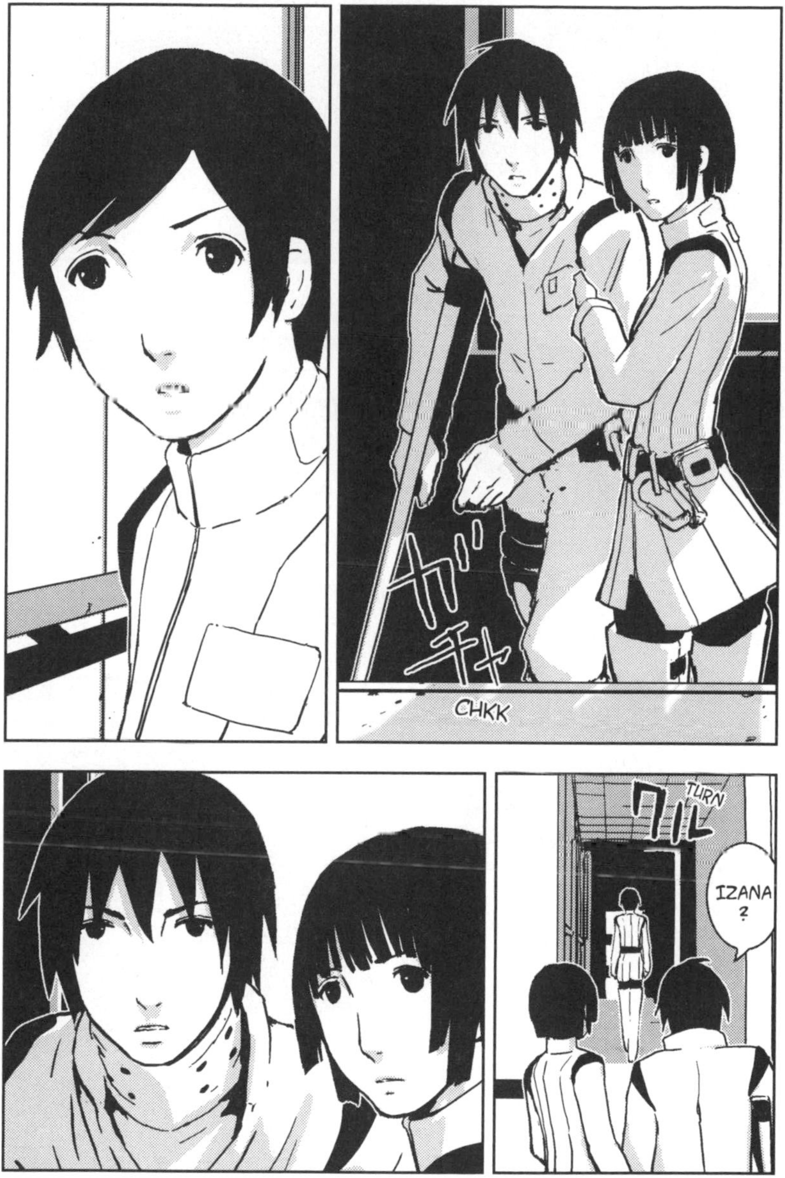
CHKK
TURN
IZANA ?

YOU CAN WALK NORMALLY ALREADY ?!
HASN'T IT ONLY BEEN THREE DAYS SINCE YOU BROKE YOUR LEG ?
OH, RIGHT. I FORGOT MY CRUTCH!
I ACTUALLY THINK I'M ALL HEALED UP,
BUT THEY TOLD ME NOT TO TAKE OFF MY CASTS ...
ゴキ CRACK
GAH... AND IT'S ITCHY ...
ゴキ CRACK
...
HUH? TODAY'S SATURDAY, ISN'T IT ?
ISN'T THIS THE MEAL DAY OF THE WEEK ?
EVERYONE IS EATING AT THE GRAVITY FESTIVAL.
THERE'S TONS OF DELICIOUS FOOD. YOU SHOULD GO TOO.

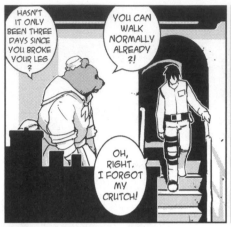

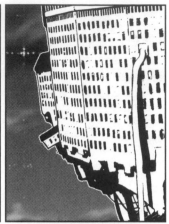

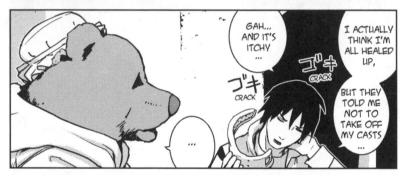

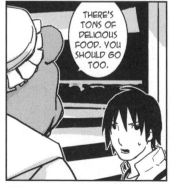

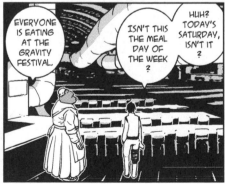

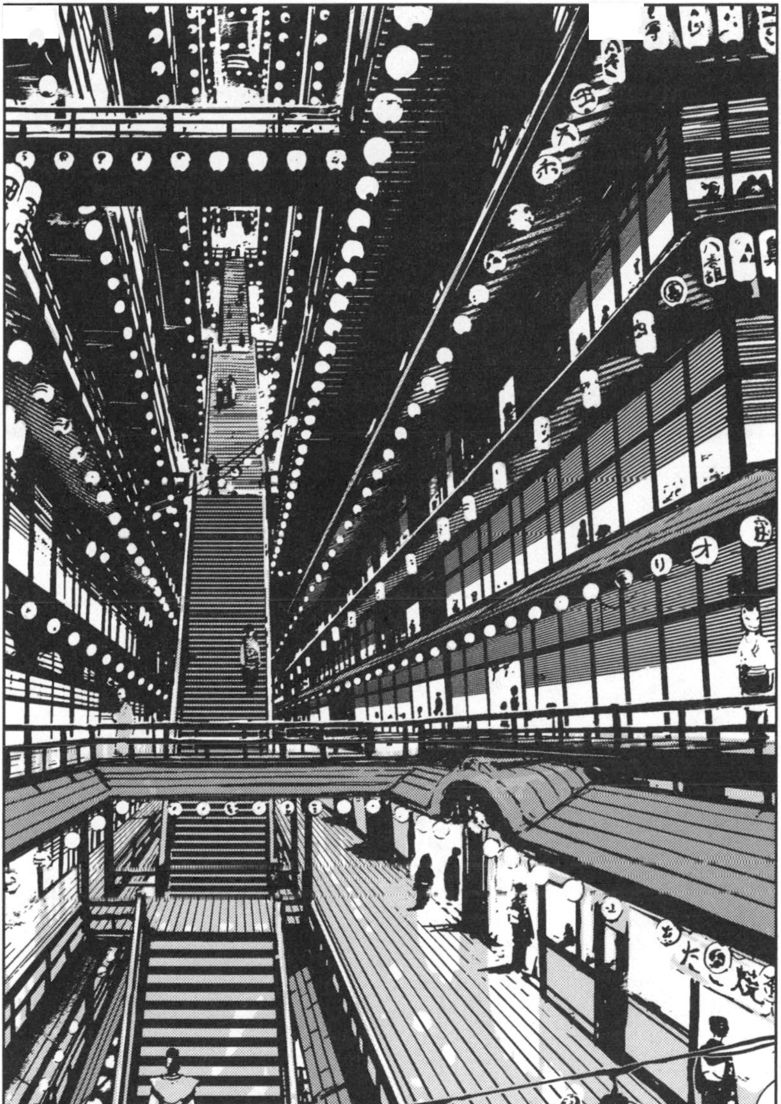

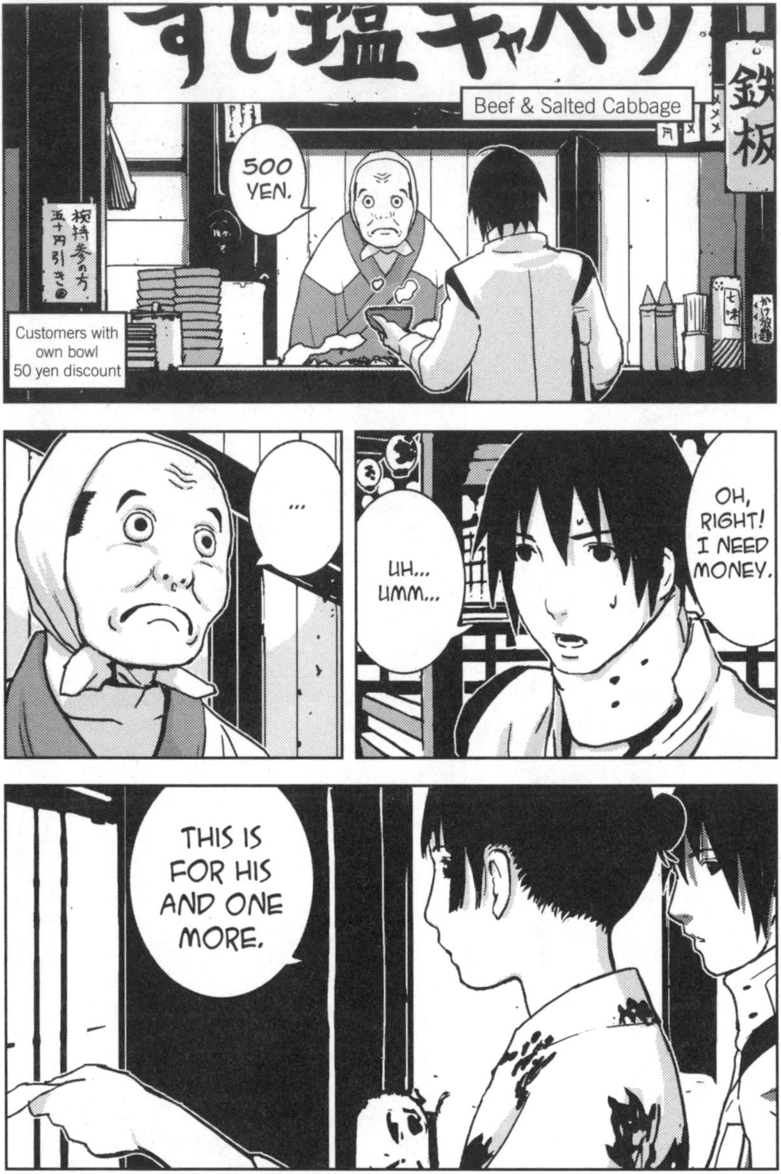
すじ塩キャベツ
鉄板
Beef & Salted Cabbage
500 YEN.
Customers with own bowl 50 yen discount
OH, RIGHT! I NEED MONEY.
UH... UMM...
...
THIS IS FOR HIS AND ONE MORE.

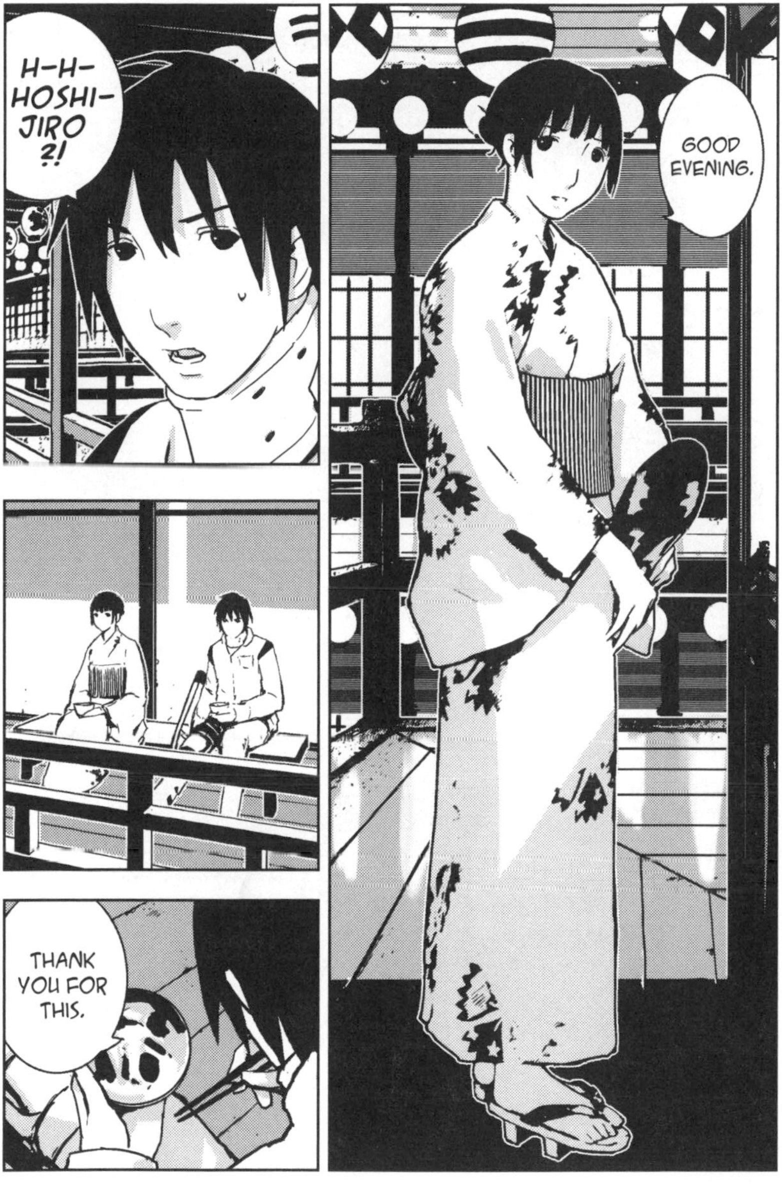
GOOD EVENING.
H-H-HOSHI-JIRO ?!
THANK YOU FOR THIS.

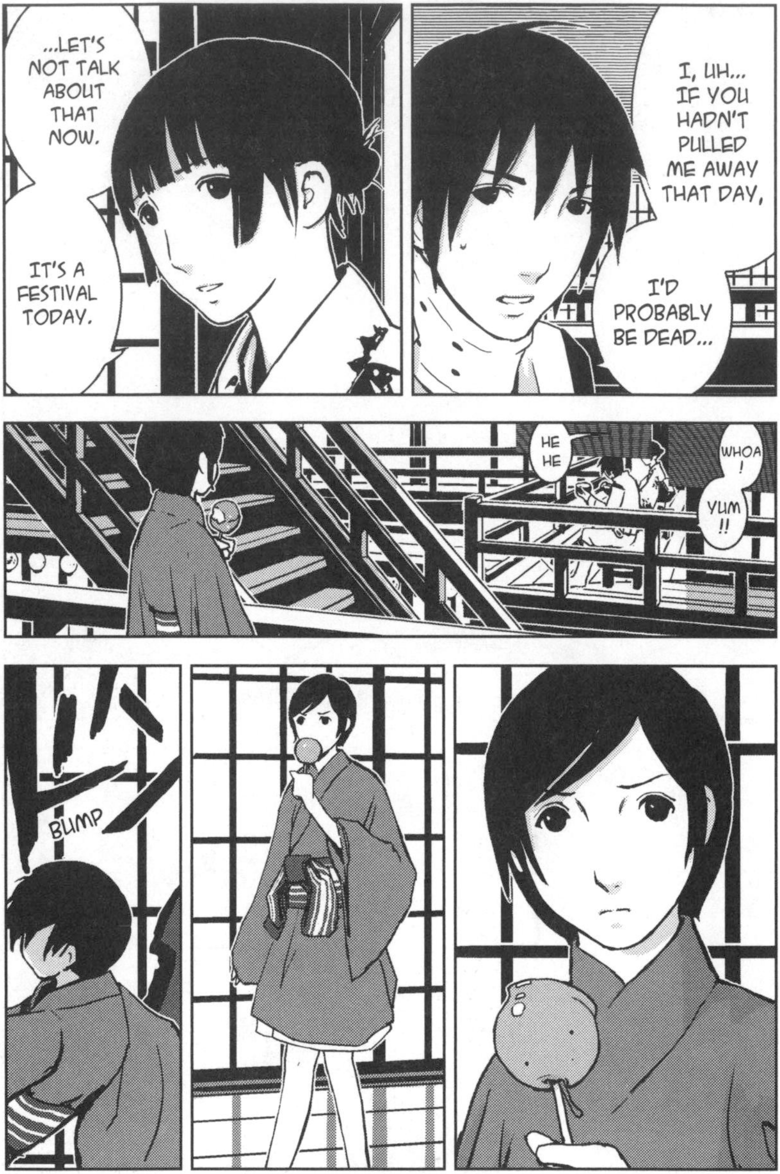
I, UH... IF YOU HADN'T PULLED ME AWAY THAT DAY,
I'D PROBABLY BE DEAD...
...LET'S NOT TALK ABOUT THAT NOW.
IT'S A FESTIVAL TODAY.
WHOA!
YUM!!
HE HE
BUMP

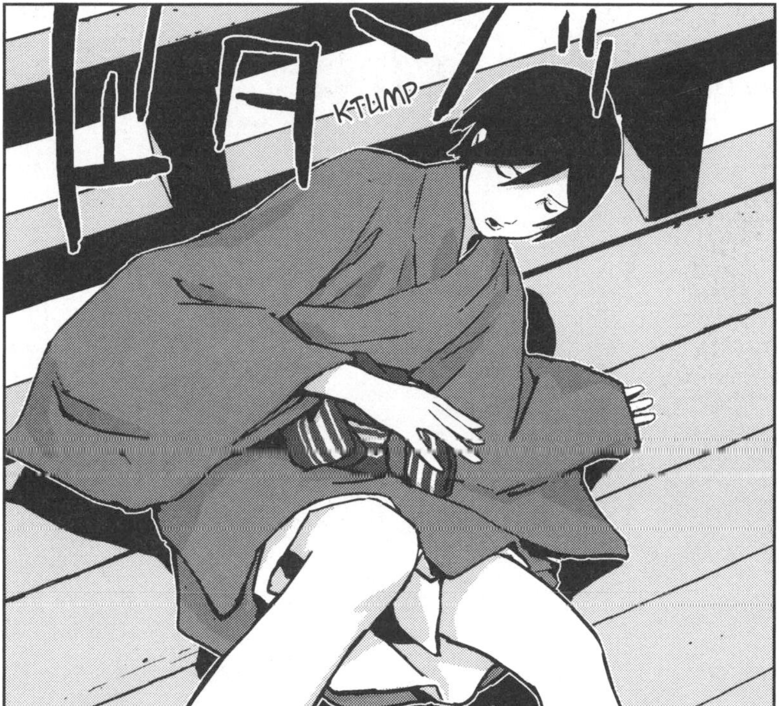
K-T-U-MP

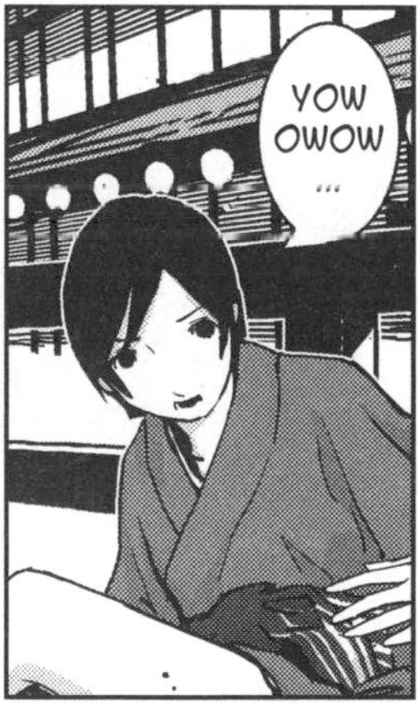
YOW OWOW ...

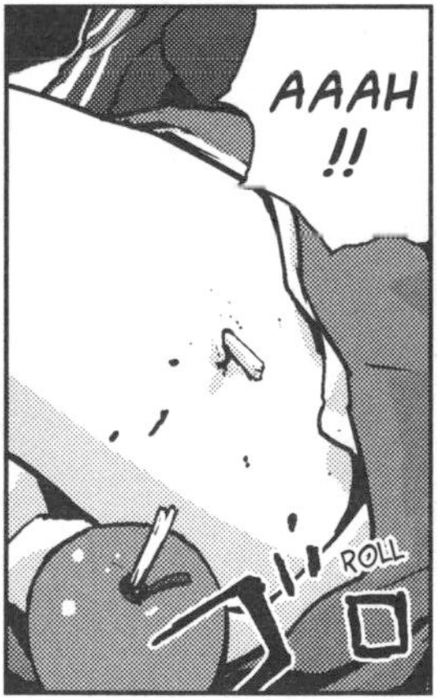
AAAH !!
ROLL

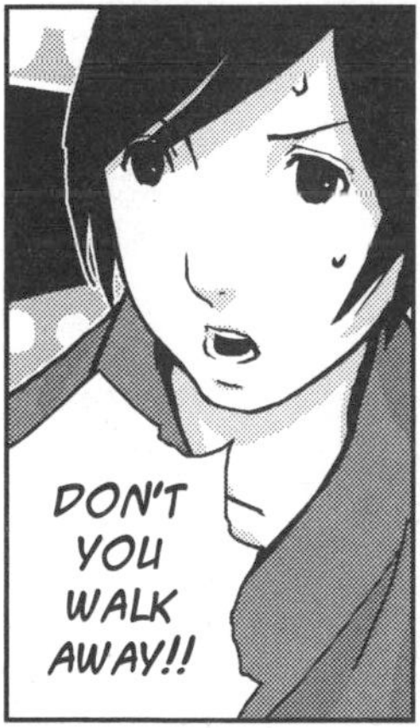
DON'T YOU WALK AWAY!!

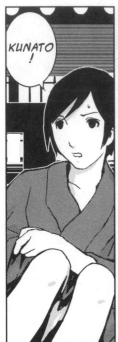

KUNATO
!

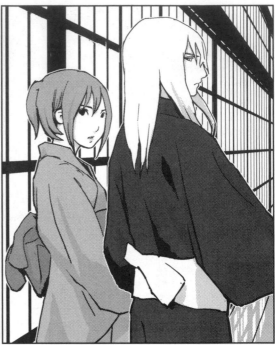

...

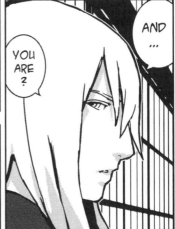

AND
...

YOU
ARE
?

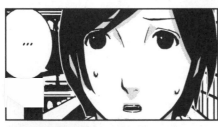

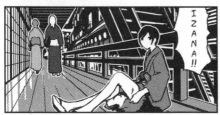

IZANA
!!

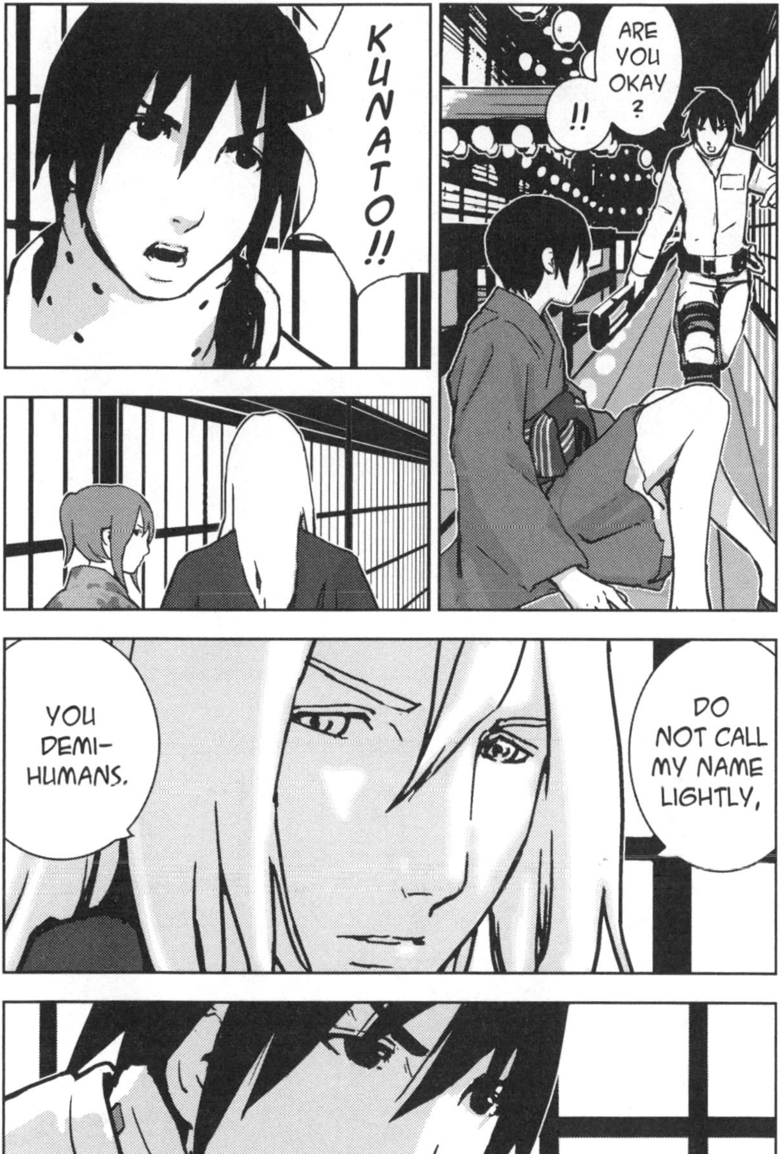
ARE YOU OKAY?
!!
KUNATO!!
DO NOT CALL MY NAME LIGHTLY,
YOU DEMI-HUMANS.

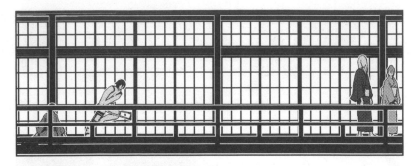

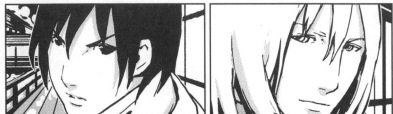

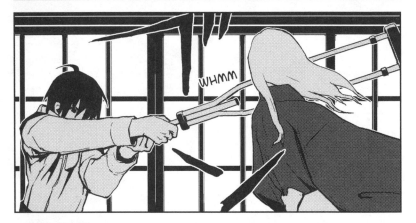

WHMM

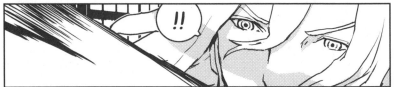

!!

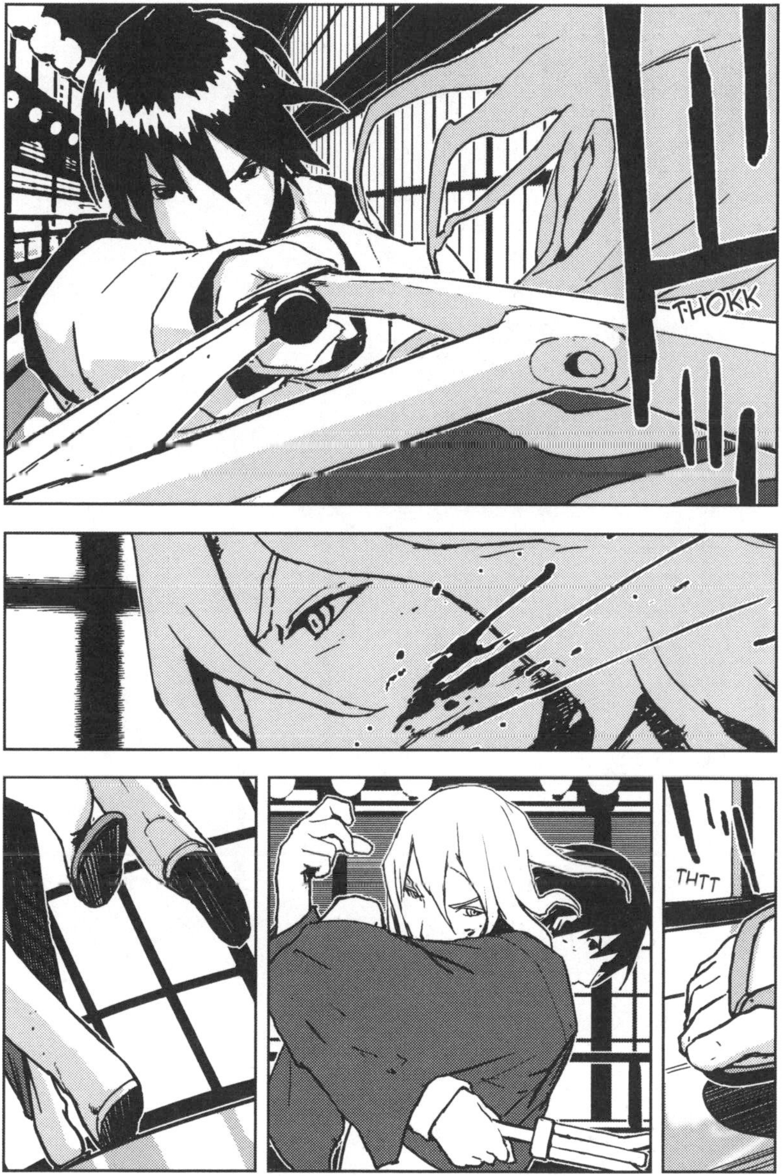
THOKK
THTT

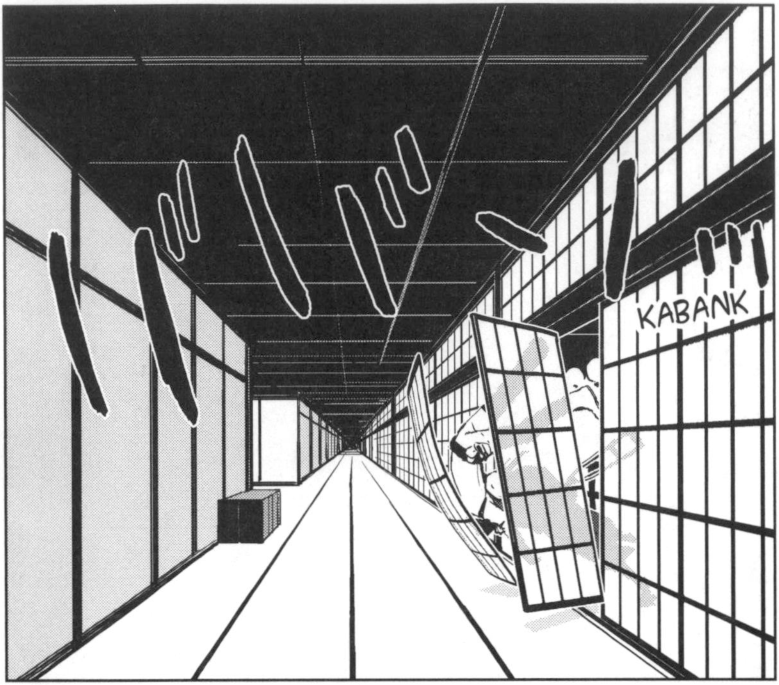
KABANK
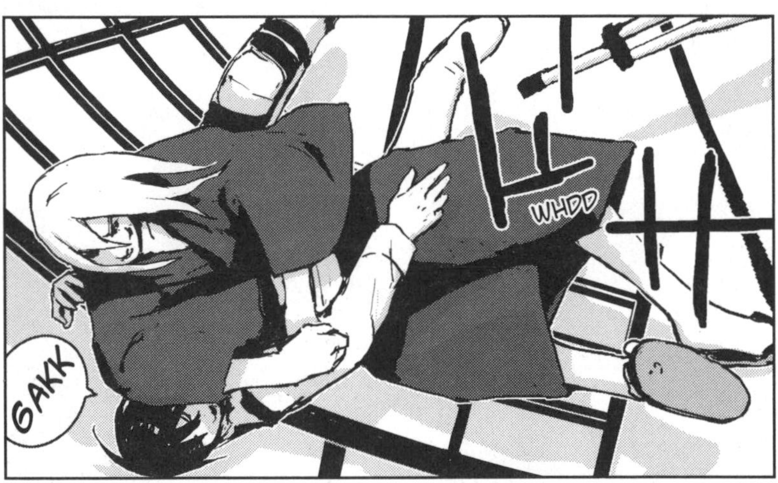
WHDD
GAKK

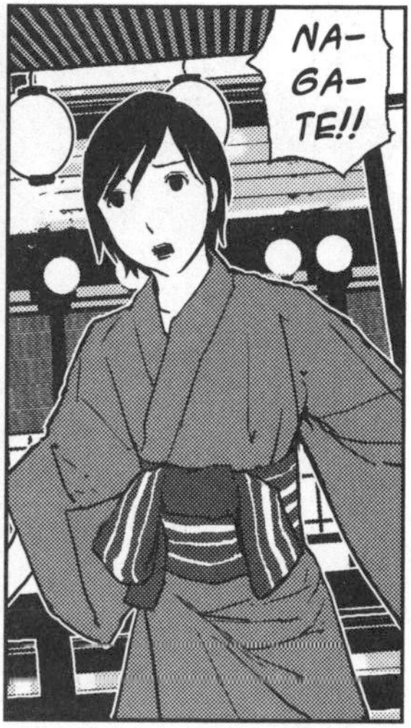
NA-
GA-
TE!!
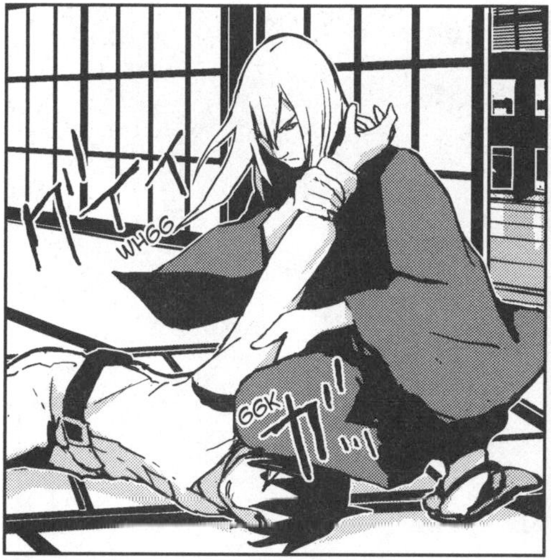
WHGG
GGK
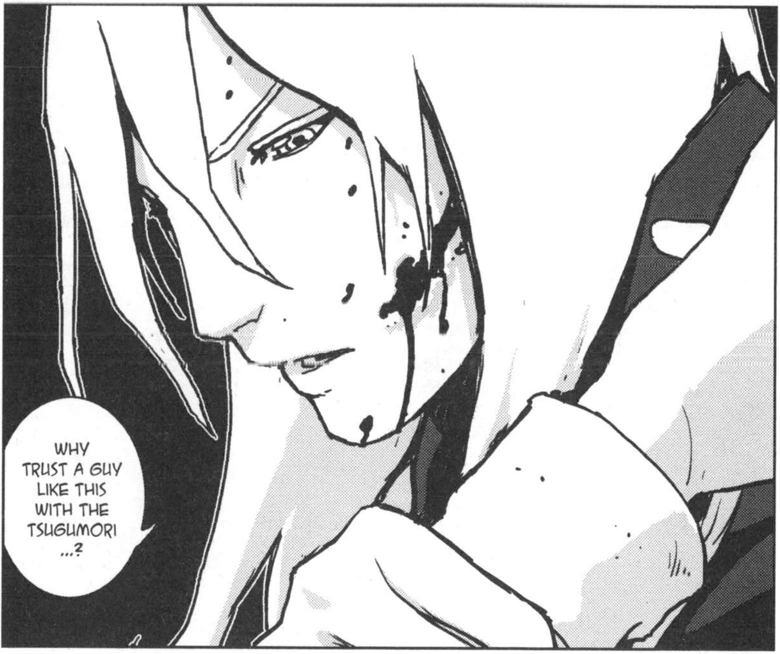
WHY TRUST A GUY LIKE THIS WITH THE TSUGUMORI ...?

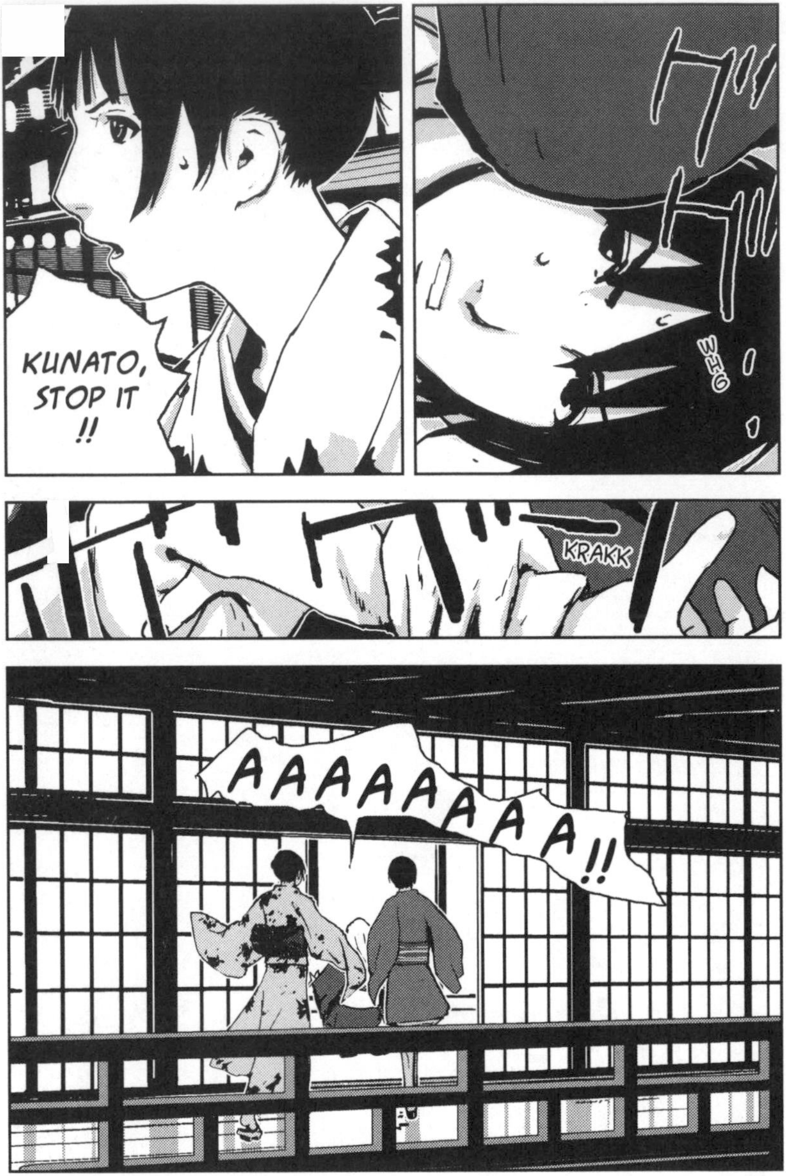
WHG
KUNATO, STOP IT !!
KRAKK
AAAAAAAA!!

Chapter 4: END

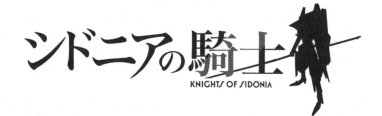

シドニアの騎士
KNIGHTS OF SIDONIA

Chapter 5: Mochikuni Akai's Glory

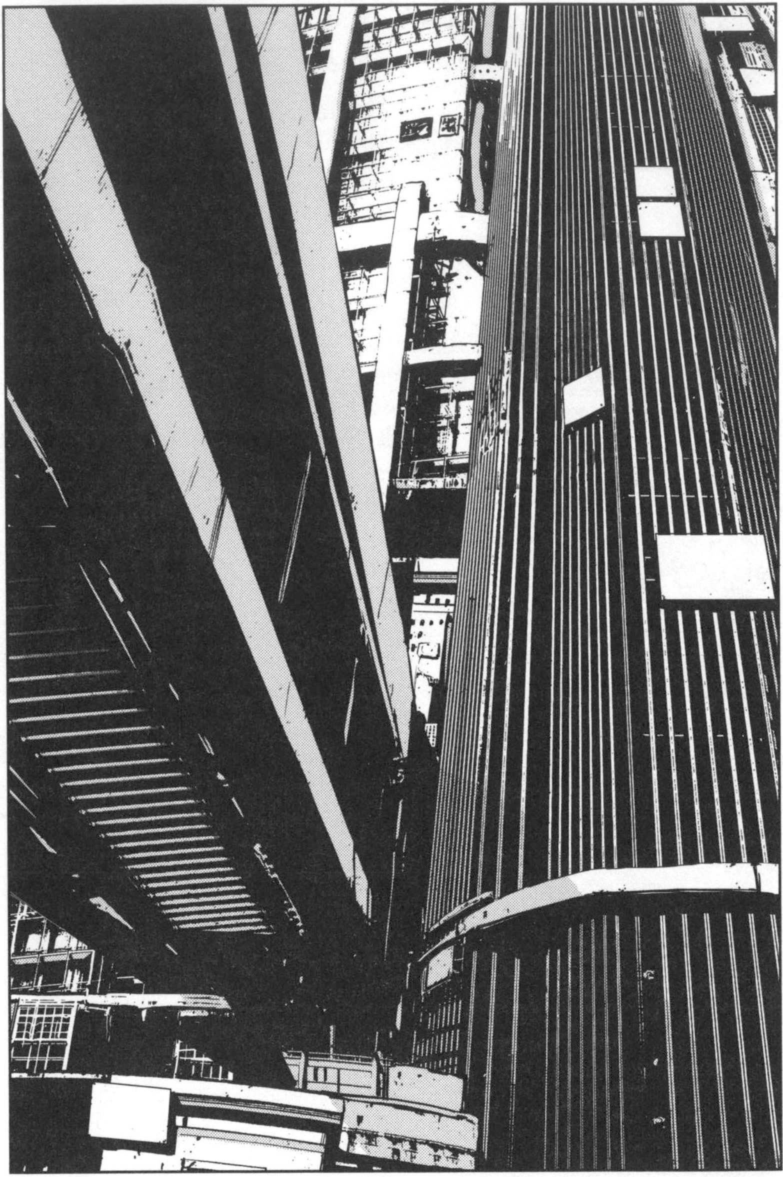

One Hundred Sights of Sidonia Part Four:
Environs of Perimeter Connection Bridge No. 16 Base, Residence Side

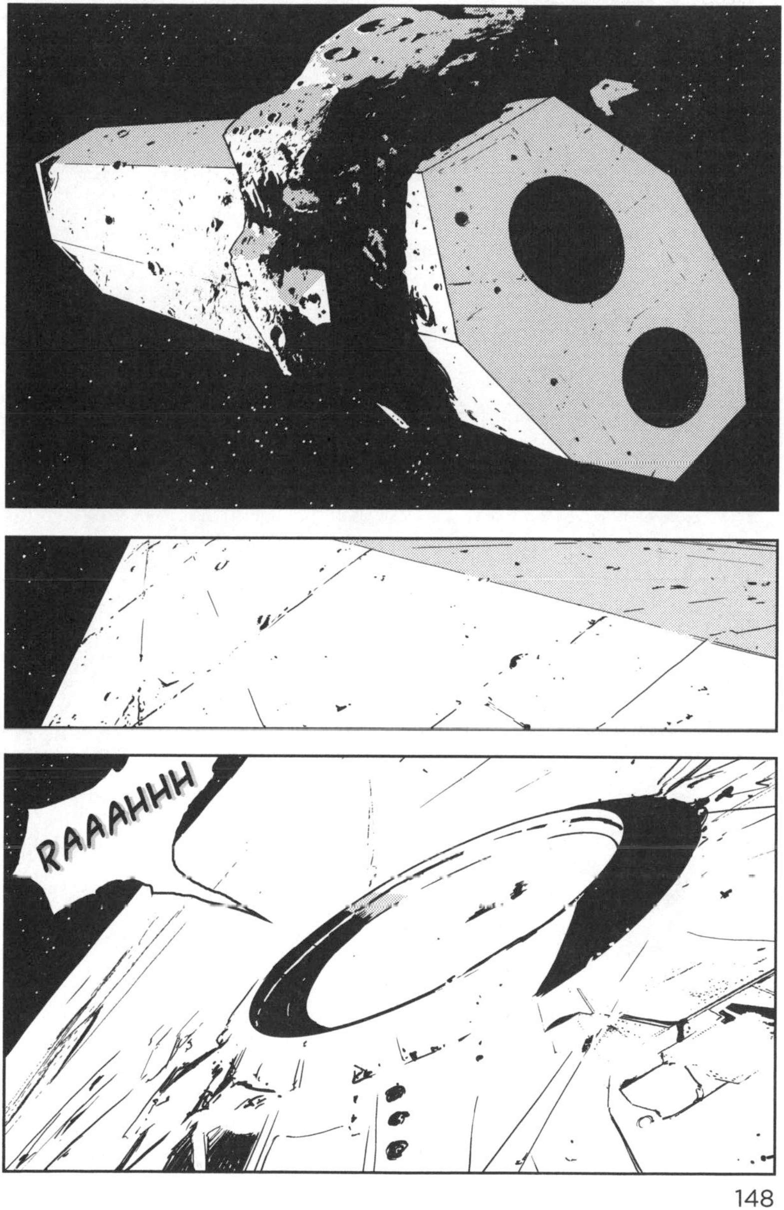
RAAAHHH

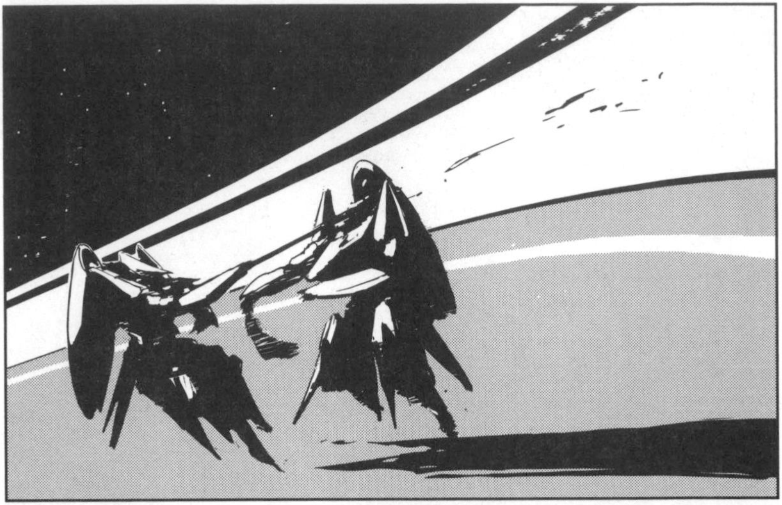

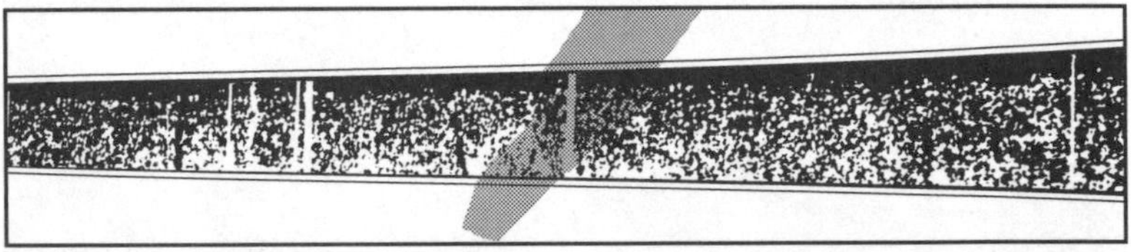

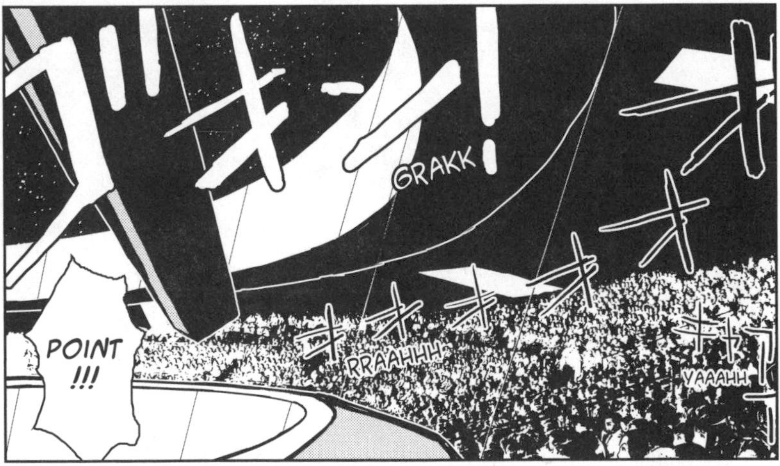
GRAKK
YAAAHH
RRAAHHH
POINT !!!

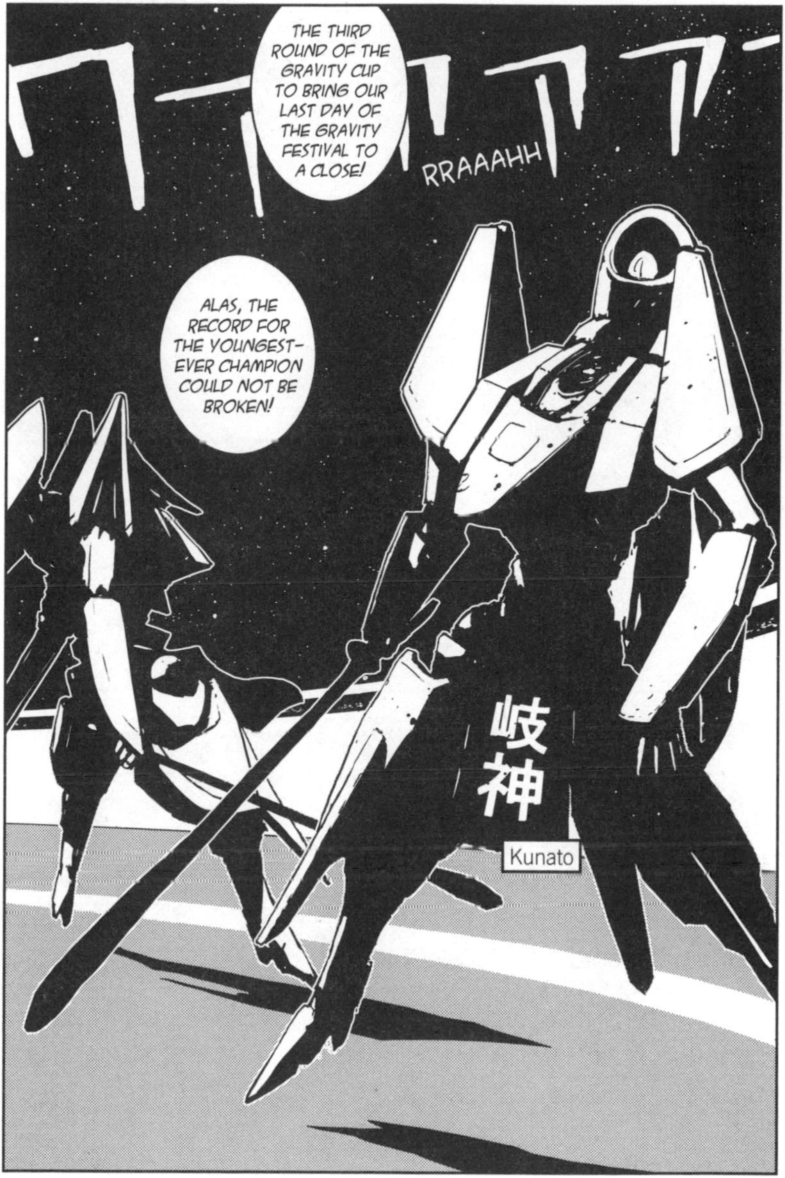
THE THIRD ROUND OF THE GRAVITY CUP TO BRING OUR LAST DAY OF THE GRAVITY FESTIVAL TO A CLOSE!
RRAAAHH
ALAS, THE RECORD FOR THE YOUNGEST-EVER CHAMPION COULD NOT BE BROKEN!
岐神
Kunato

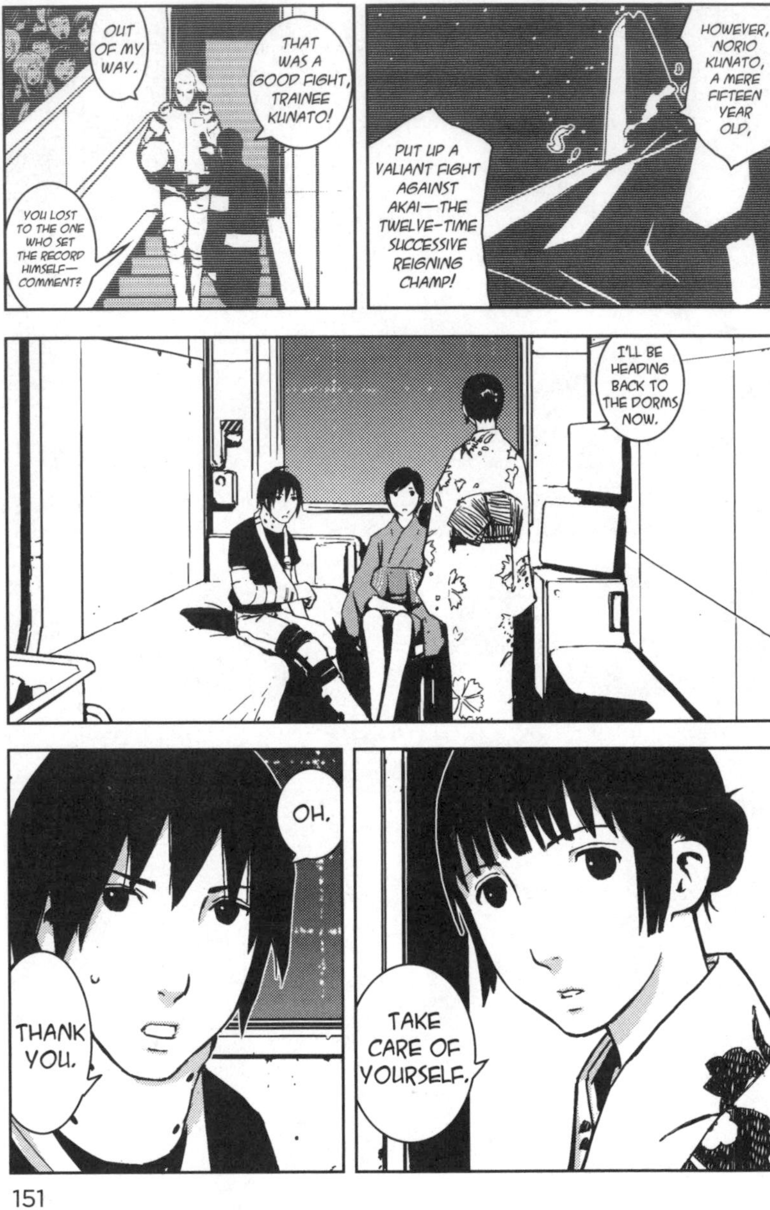

HOWEVER, NORIO KUNATO, A MERE FIFTEEN YEAR OLD,
PUT UP A VALIANT FIGHT AGAINST AKAI—THE TWELVE-TIME SUCCESSIVE REIGNING CHAMP!
THAT WAS A GOOD FIGHT, TRAINEE KUNATO!
OUT OF MY WAY.
YOU LOST TO THE ONE WHO SET THE RECORD HIMSELF—COMMENT?
I'LL BE HEADING BACK TO THE DORMS NOW.
TAKE CARE OF YOURSELF.
OH.
THANK YOU.

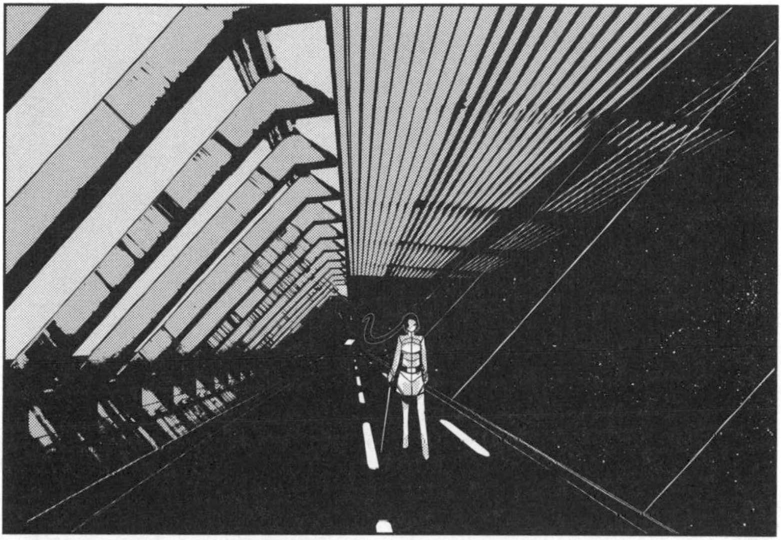

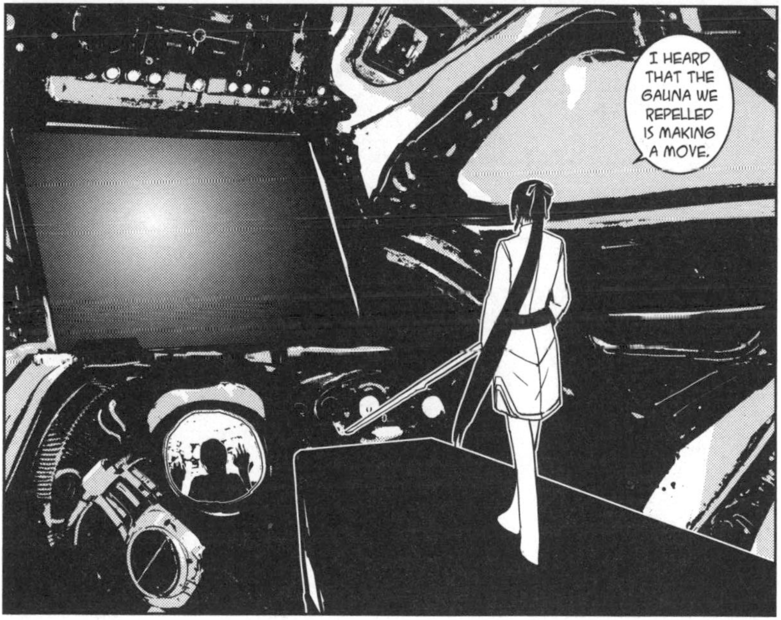
I HEARD THAT THE GAUNA WE REPELLED IS MAKING A MOVE.

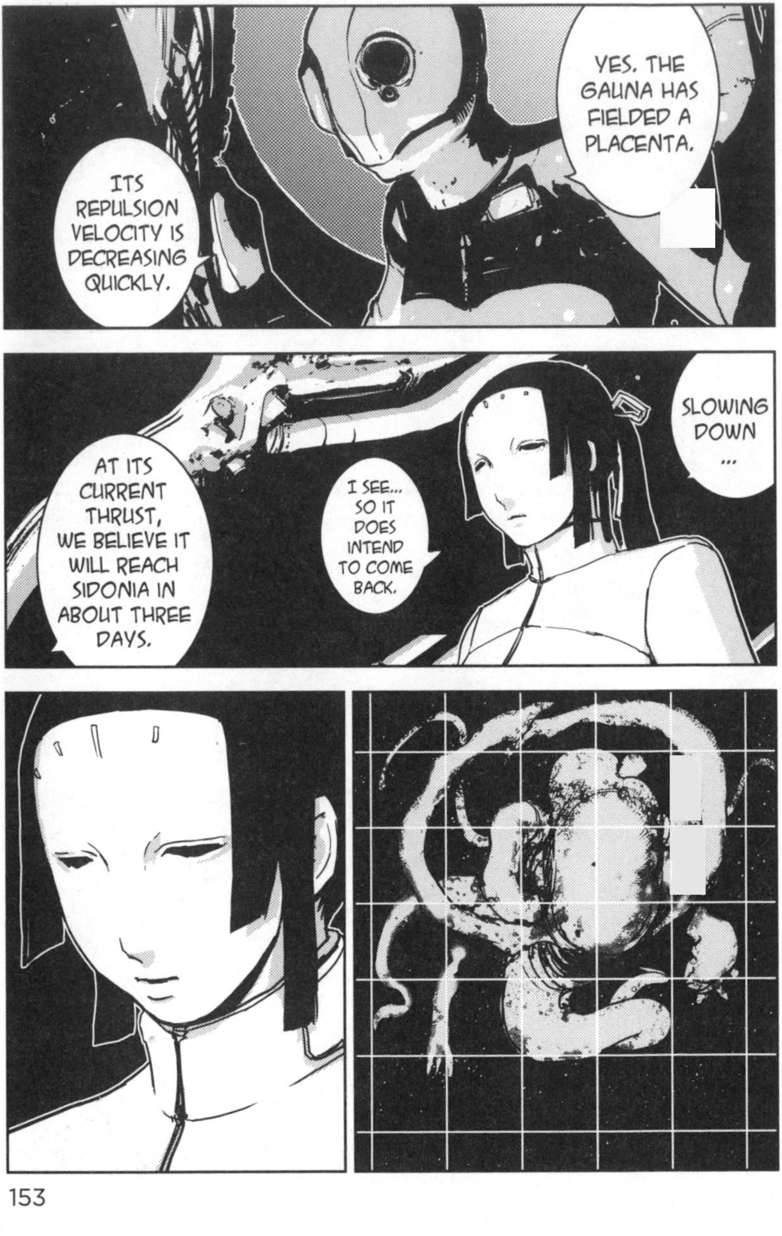
YES. THE GAUNA HAS FIELDED A PLACENTA.
ITS REPULSION VELOCITY IS DECREASING QUICKLY.
SLOWING DOWN ...
I SEE... SO IT DOES INTEND TO COME BACK.
AT ITS CURRENT THRUST, WE BELIEVE IT WILL REACH SIDONIA IN ABOUT THREE DAYS.

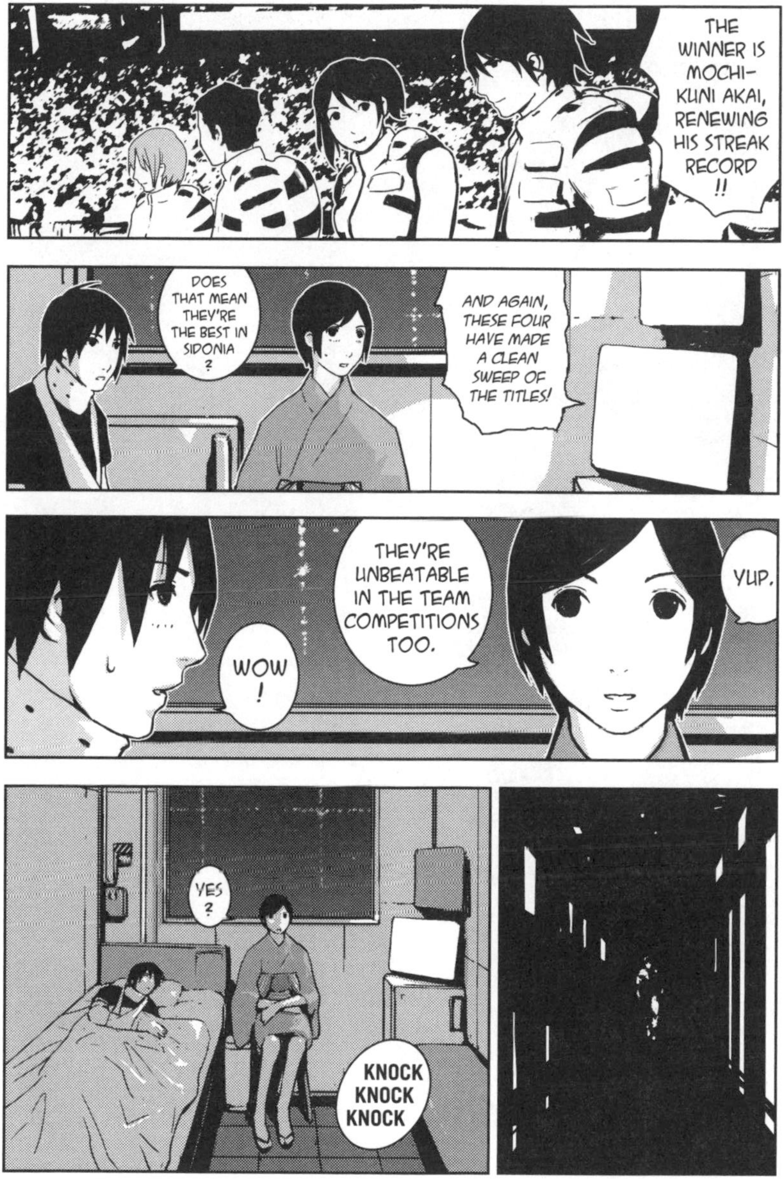
THE WINNER IS MOCHI-KUNI AKAI, RENEWING HIS STREAK RECORD !!
AND AGAIN, THESE FOUR HAVE MADE A CLEAN SWEEP OF THE TITLES!
DOES THAT MEAN THEY'RE THE BEST IN SIDONIA ?
YUP.
THEY'RE UNBEATABLE IN THE TEAM COMPETITIONS TOO.
...
WOW !
YES ?
KNOCK KNOCK KNOCK

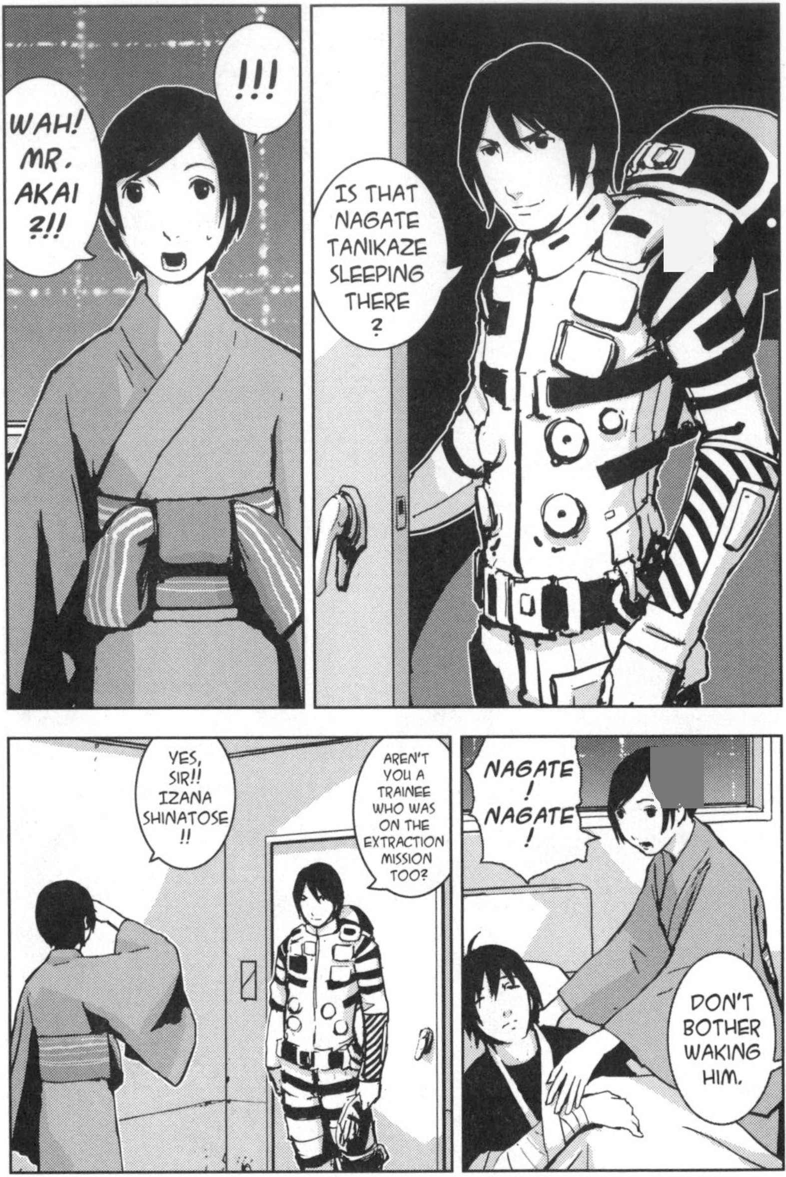
IS THAT NAGATE TANIKAZE SLEEPING THERE?
!!!
WAH! MR. AKAI?!!
NAGATE! NAGATE!
DON'T BOTHER WAKING HIM.
AREN'T YOU A TRAINEE WHO WAS ON THE EXTRACTION MISSION TOO?
YES, SIR!! IZANA SHINATOSE!!

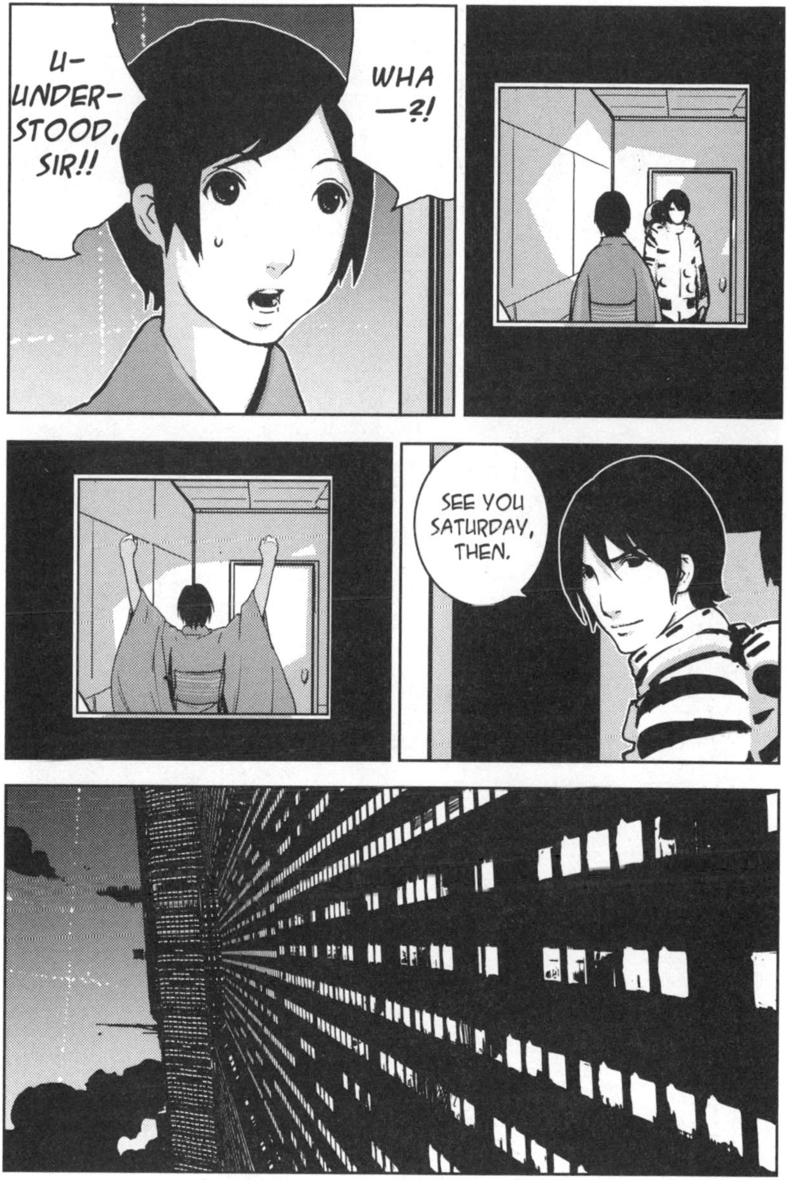
WHA—?!
U-UNDERSTOOD, SIR!!
SEE YOU SATURDAY, THEN.

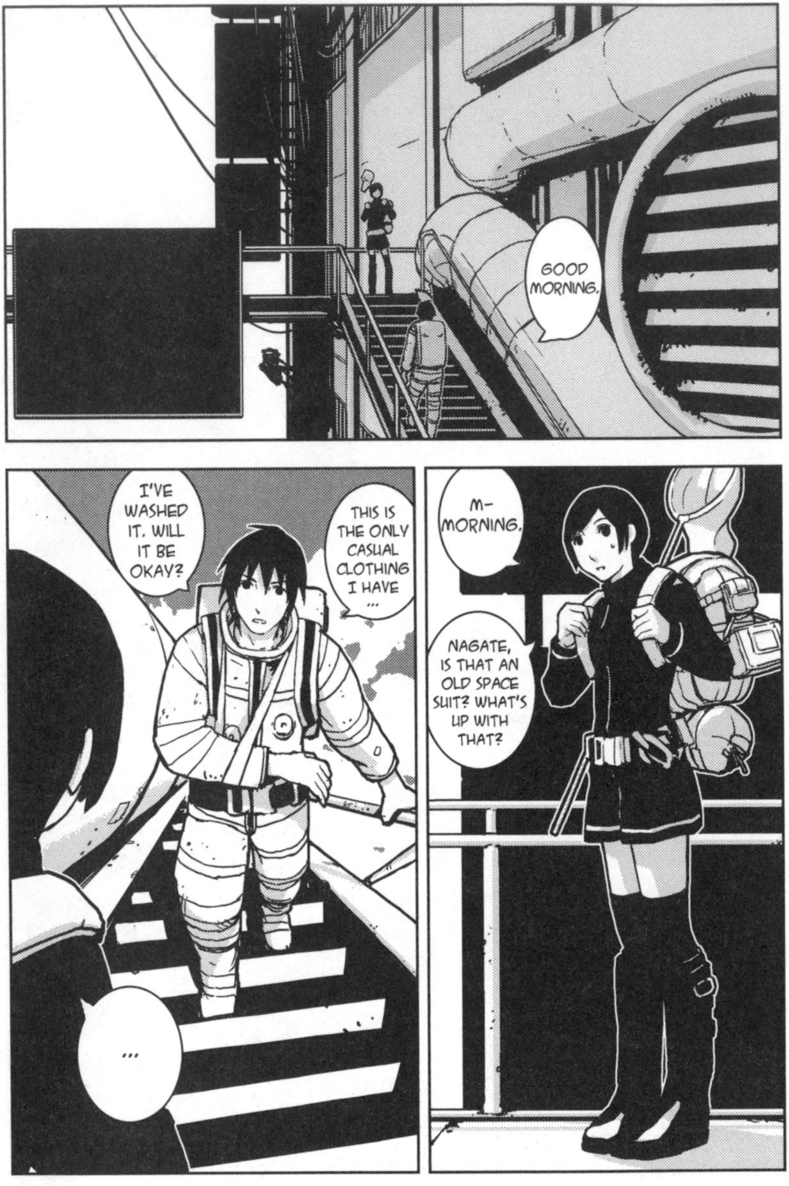
GOOD MORNING.
M-MORNING.
NAGATE, IS THAT AN OLD SPACE SUIT? WHAT'S UP WITH THAT?
THIS IS THE ONLY CASUAL CLOTHING I HAVE ...
I'VE WASHED IT. WILL IT BE OKAY?
...

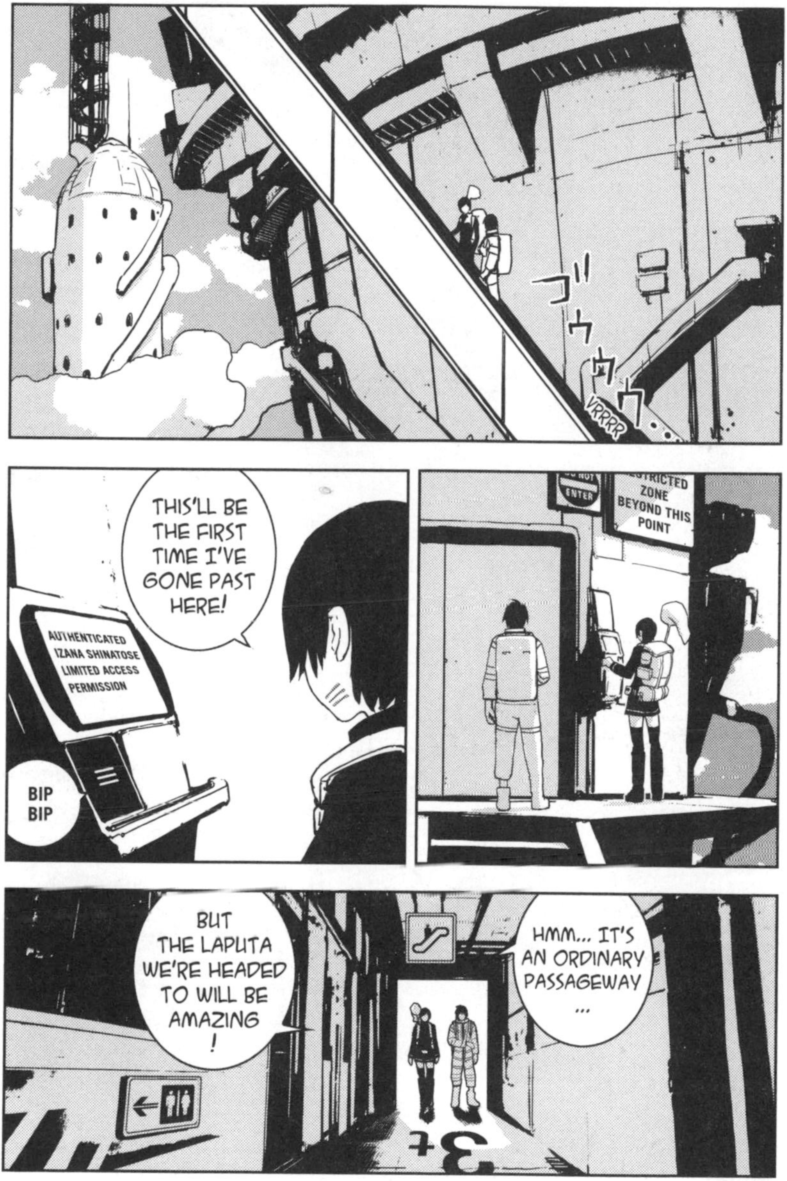
ゴウウウ
VRRRR
ENTER
STRICTED ZONE BEYOND THIS POINT
AUTHENTICATED IZANA SHINATOSE LIMITED ACCESS PERMISSION
THIS'LL BE THE FIRST TIME I'VE GONE PAST HERE!
BIP BIP
HMM... IT'S AN ORDINARY PASSAGEWAY ...
BUT THE LAPUTA WE'RE HEADED TO WILL BE AMAZING !
3F

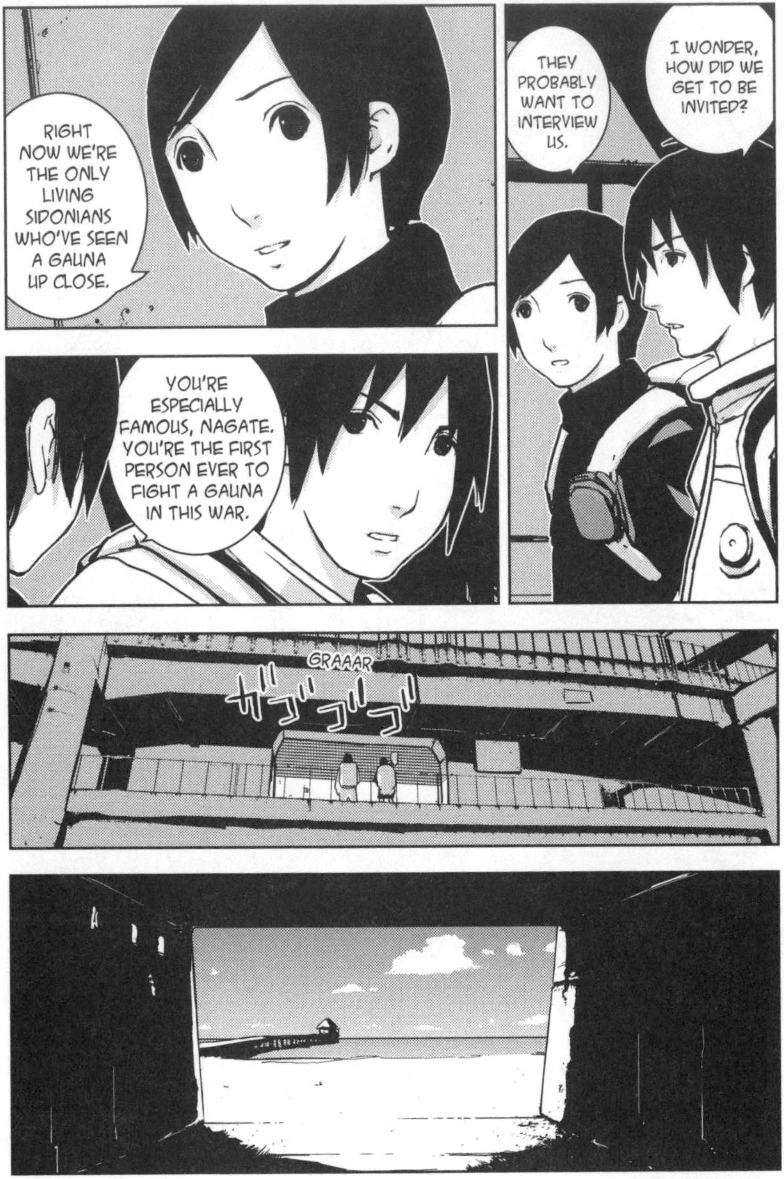
I WONDER, HOW DID WE GET TO BE INVITED?
THEY PROBABLY WANT TO INTERVIEW US.
RIGHT NOW WE'RE THE ONLY LIVING SIDONIANS WHO'VE SEEN A GAUNA UP CLOSE.
YOU'RE ESPECIALLY FAMOUS, NAGATE. YOU'RE THE FIRST PERSON EVER TO FIGHT A GAUNA IN THIS WAR.
GRAAAR
ガゴゴゴ

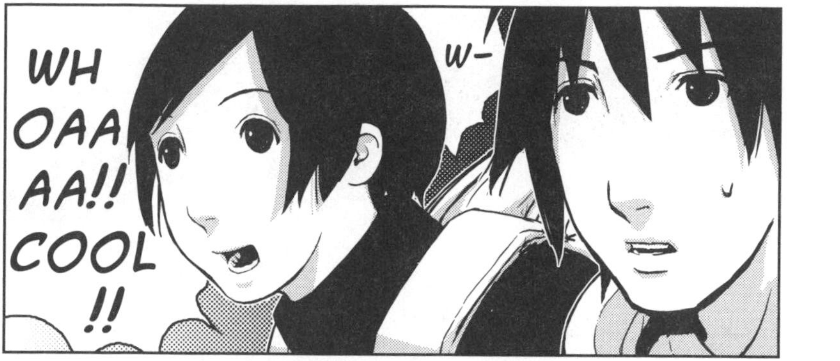
W-
WH
OAA
AA!!
COOL
!!

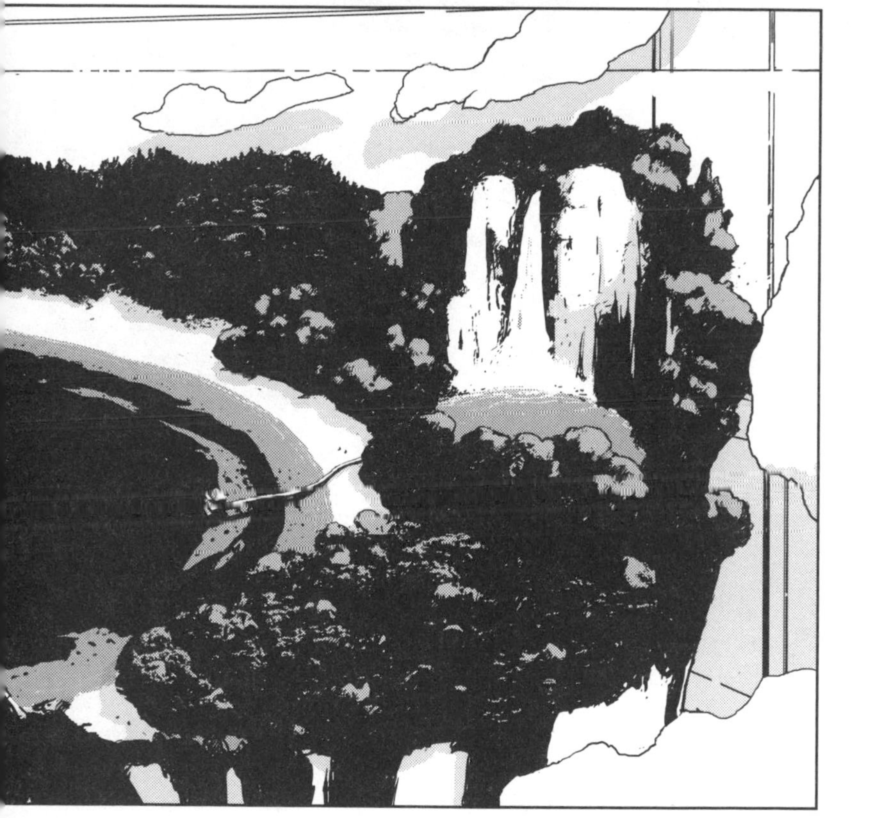

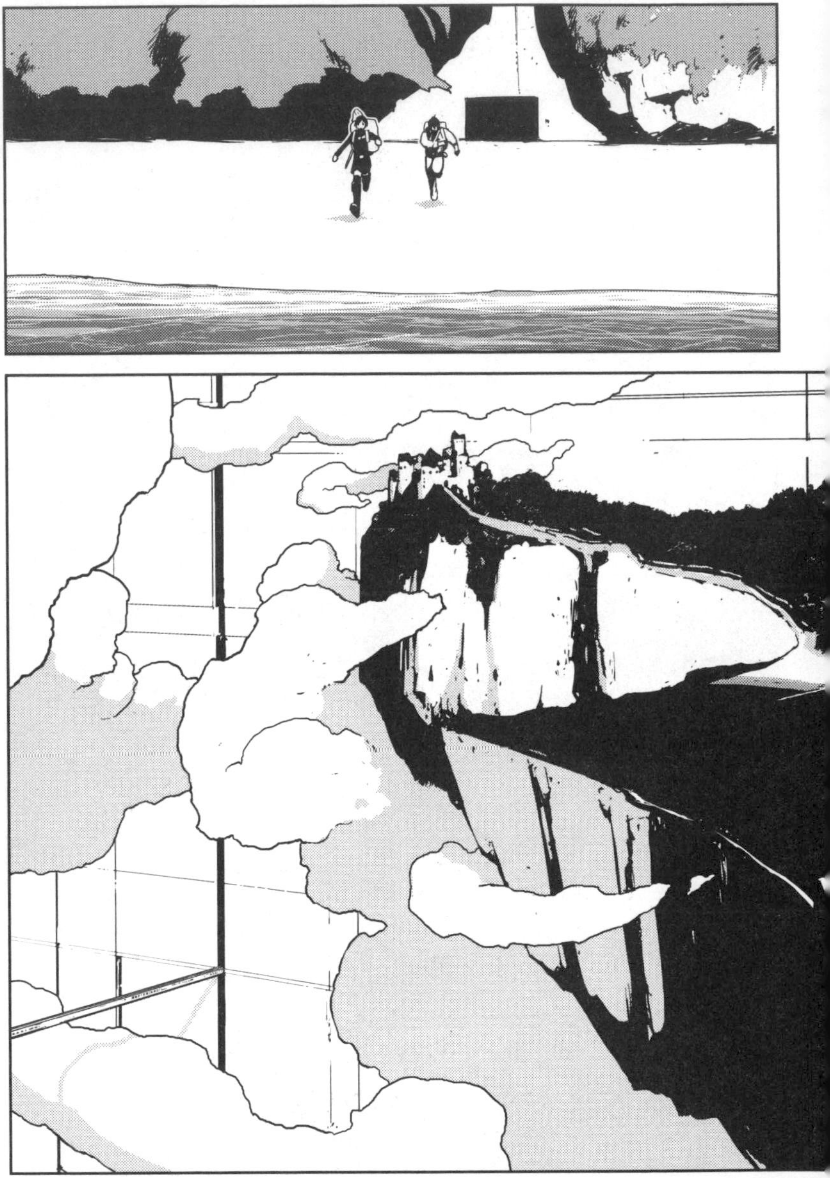

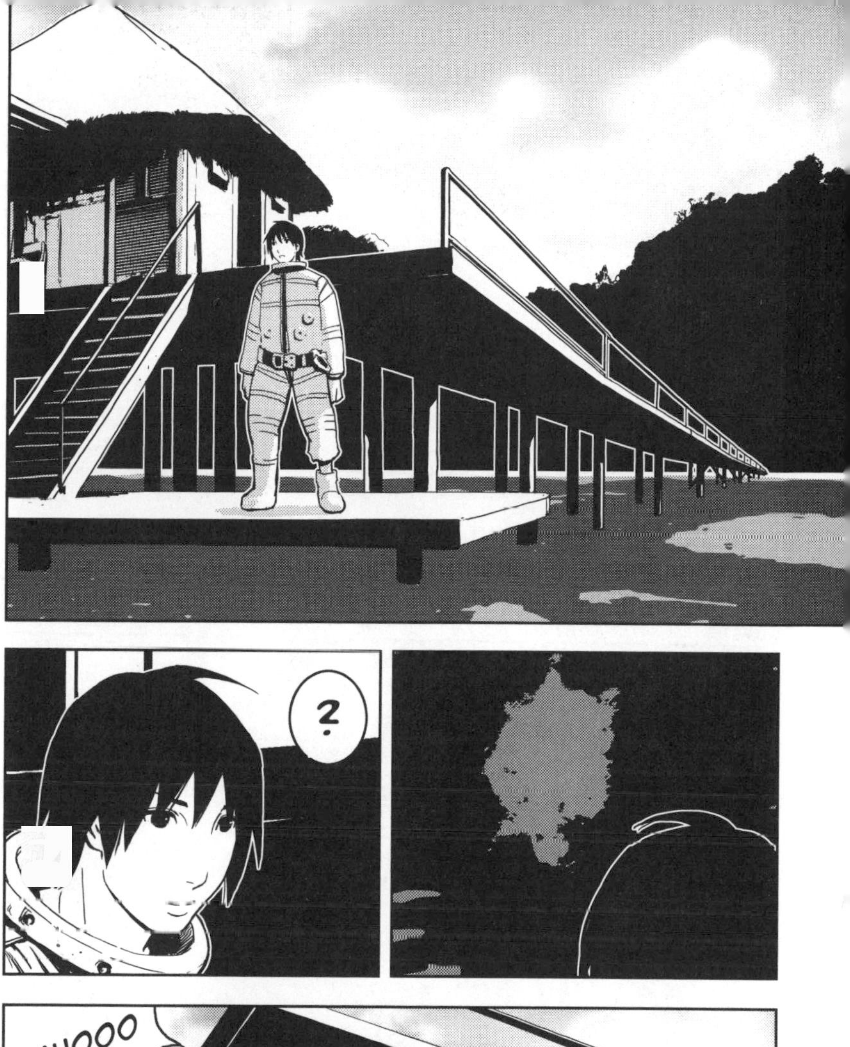
?
YAHOOO !!
TMP TMP TMP TMP TMP TMP TMP

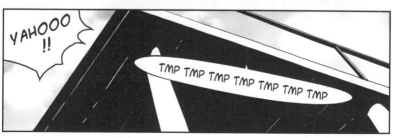

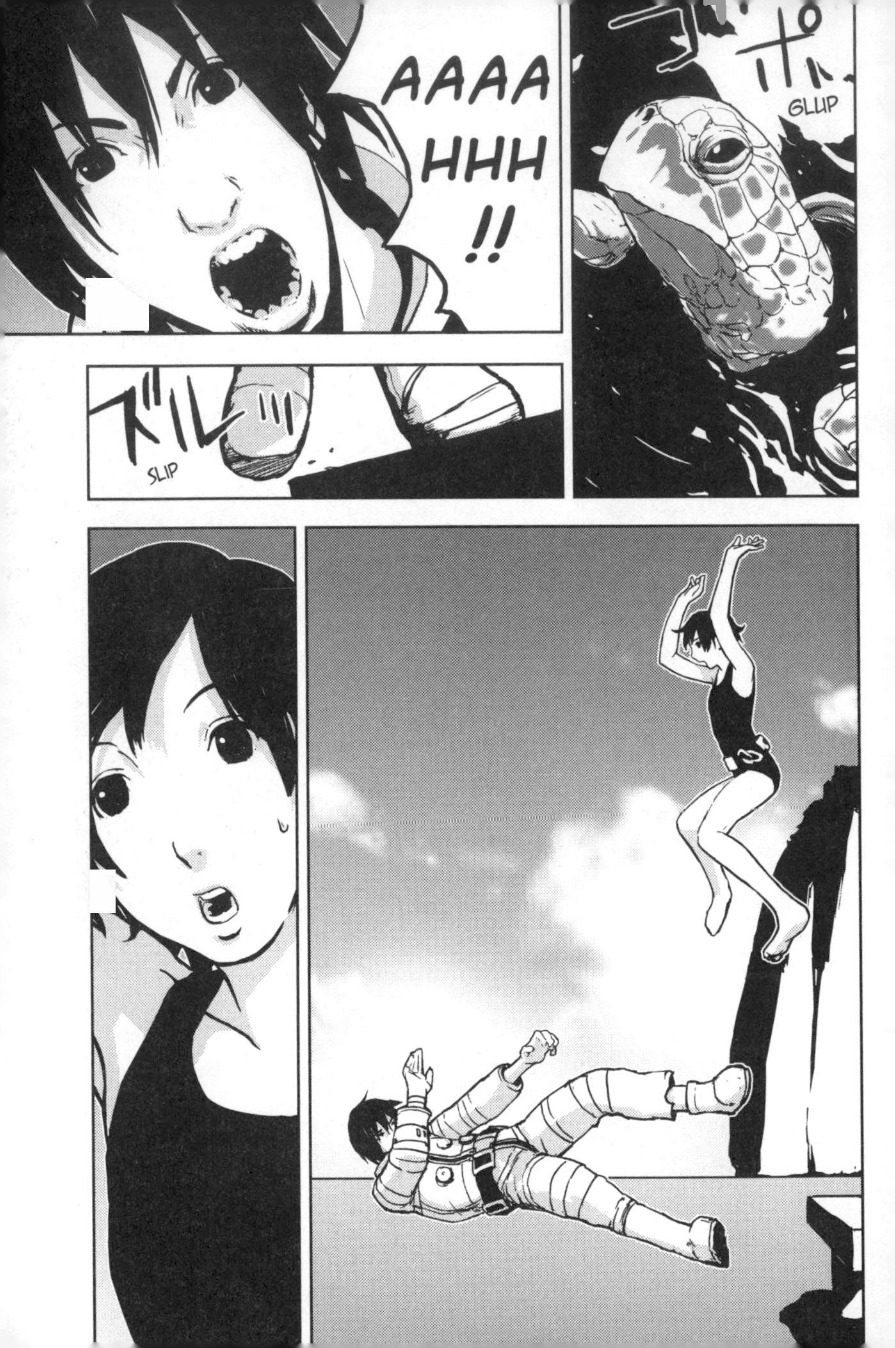
GLUP
AAAAHHH!!
SLIP

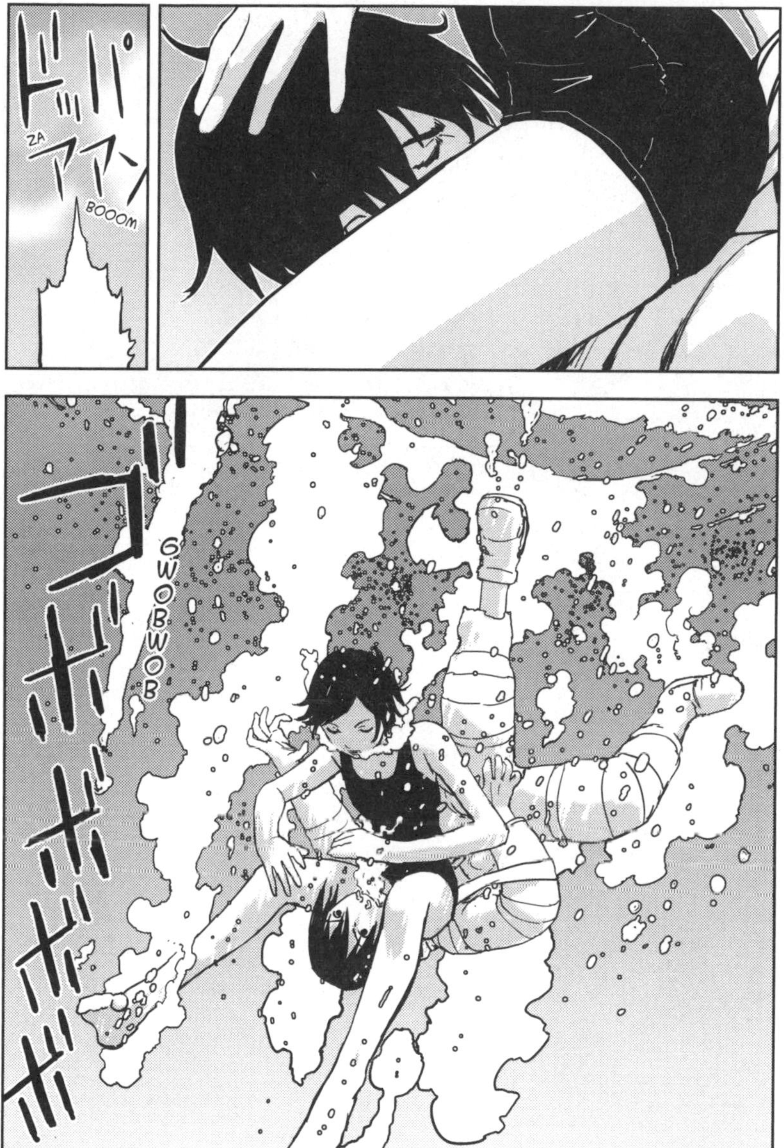
ドッパアアン
ZA
BOOOM
ゴボボボ
GWOBWOBWOB

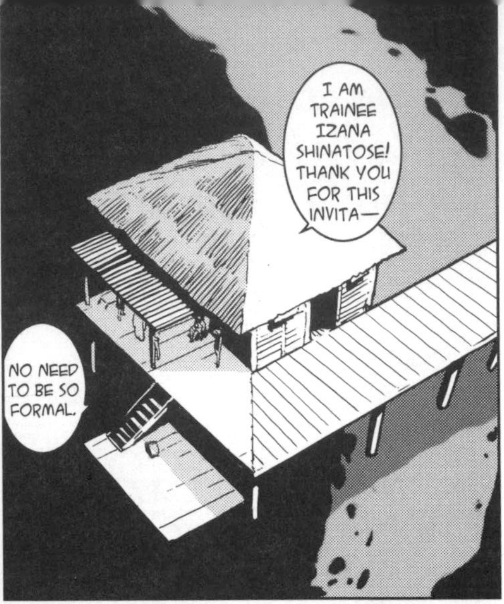
I AM TRAINEE IZANA SHINATOSE! THANK YOU FOR THIS INVITA—
NO NEED TO BE SO FORMAL.

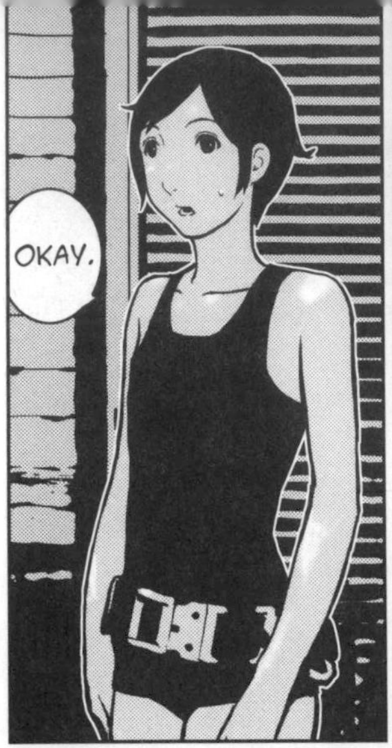
OKAY.

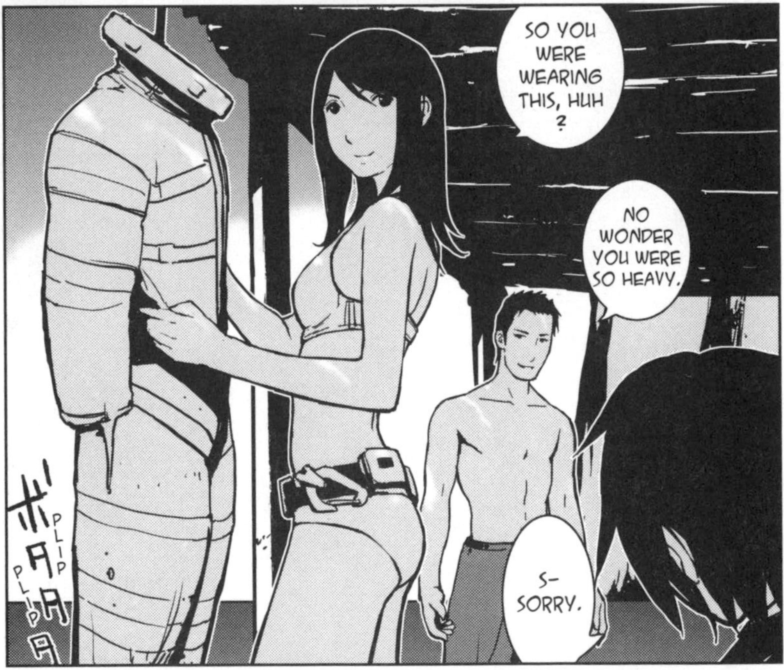
SO YOU WERE WEARING THIS, HUH?
NO WONDER YOU WERE SO HEAVY.
S-SORRY.
PLIP
PLIP

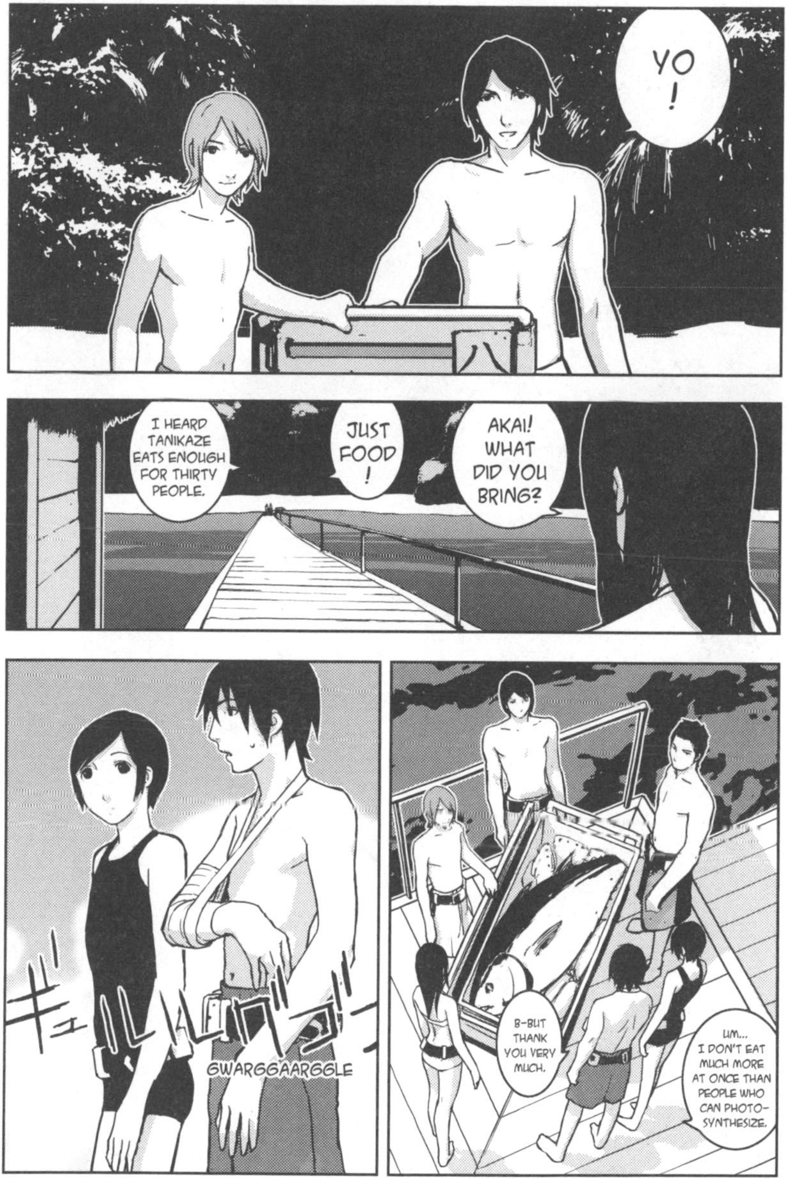
YO!
AKAI! WHAT DID YOU BRING?
JUST FOOD!
I HEARD TANIKAZE EATS ENOUGH FOR THIRTY PEOPLE.
UM... I DON'T EAT MUCH MORE AT ONCE THAN PEOPLE WHO CAN PHOTO-SYNTHESIZE.
B-BUT THANK YOU VERY MUCH.
GWARGGAARGGLE

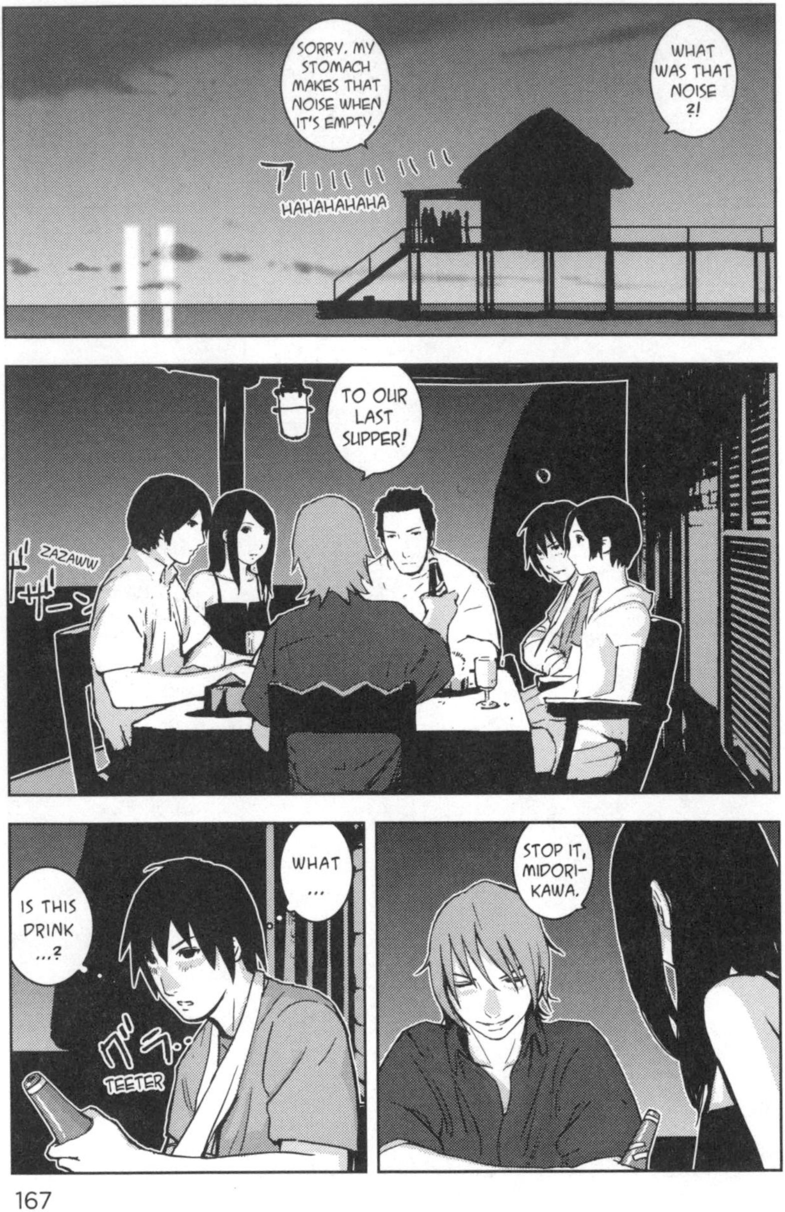
WHAT WAS THAT NOISE ?!
SORRY, MY STOMACH MAKES THAT NOISE WHEN IT'S EMPTY.
HAHAHAHAHA
TO OUR LAST SUPPER!
ZAZAWW
STOP IT, MIDORI-KAWA.
WHAT ...
IS THIS DRINK ...?
TEETER

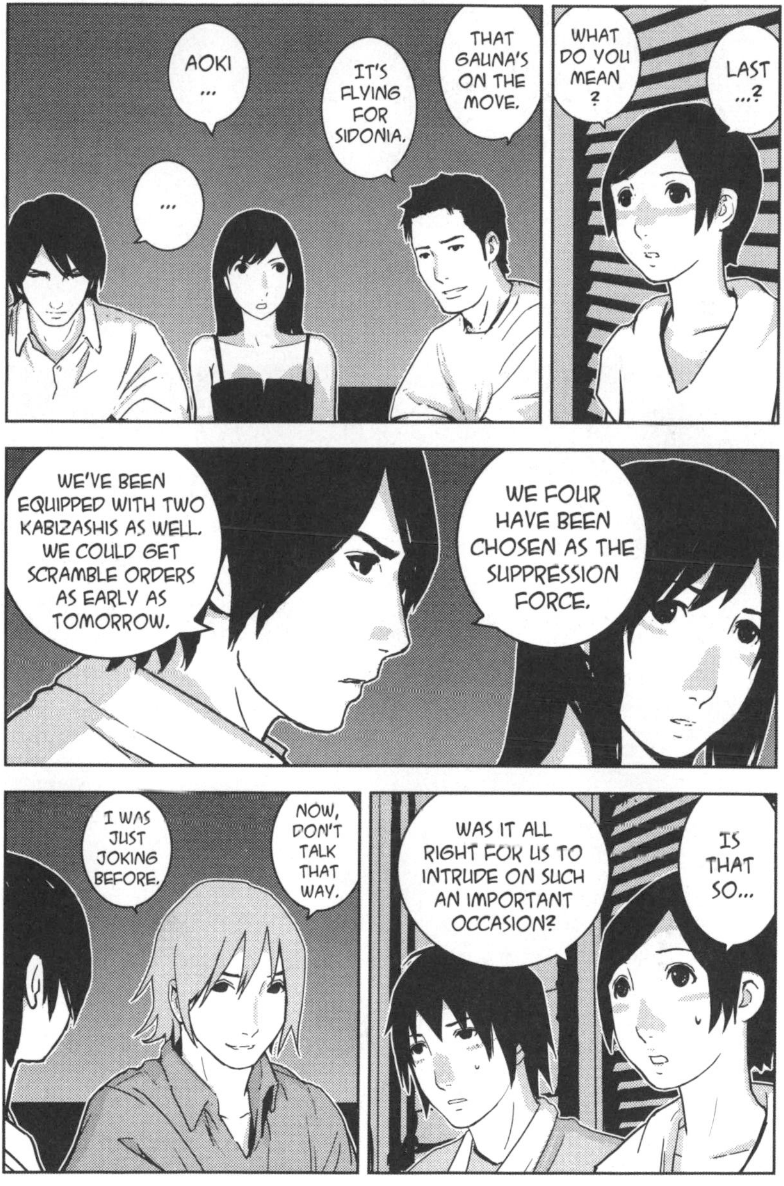
LAST ...?
WHAT DO YOU MEAN?
THAT GAUNA'S ON THE MOVE.
IT'S FLYING FOR SIDONIA.
AOKI ...
...
WE FOUR HAVE BEEN CHOSEN AS THE SUPPRESSION FORCE.
WE'VE BEEN EQUIPPED WITH TWO KABIZASHIS AS WELL. WE COULD GET SCRAMBLE ORDERS AS EARLY AS TOMORROW.
IS THAT SO...
WAS IT ALL RIGHT FOR US TO INTRUDE ON SUCH AN IMPORTANT OCCASION?
NOW, DON'T TALK THAT WAY.
I WAS JUST JOKING BEFORE.

!
BDAM!!
NAGA-TE!
ザザ・・・・ーー。。
ZAZAWW

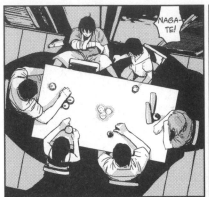

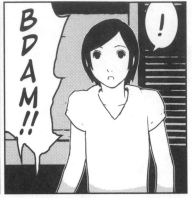

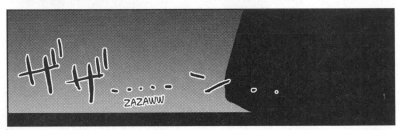

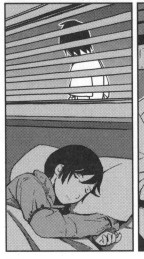

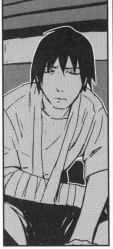

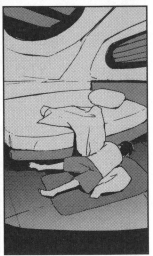

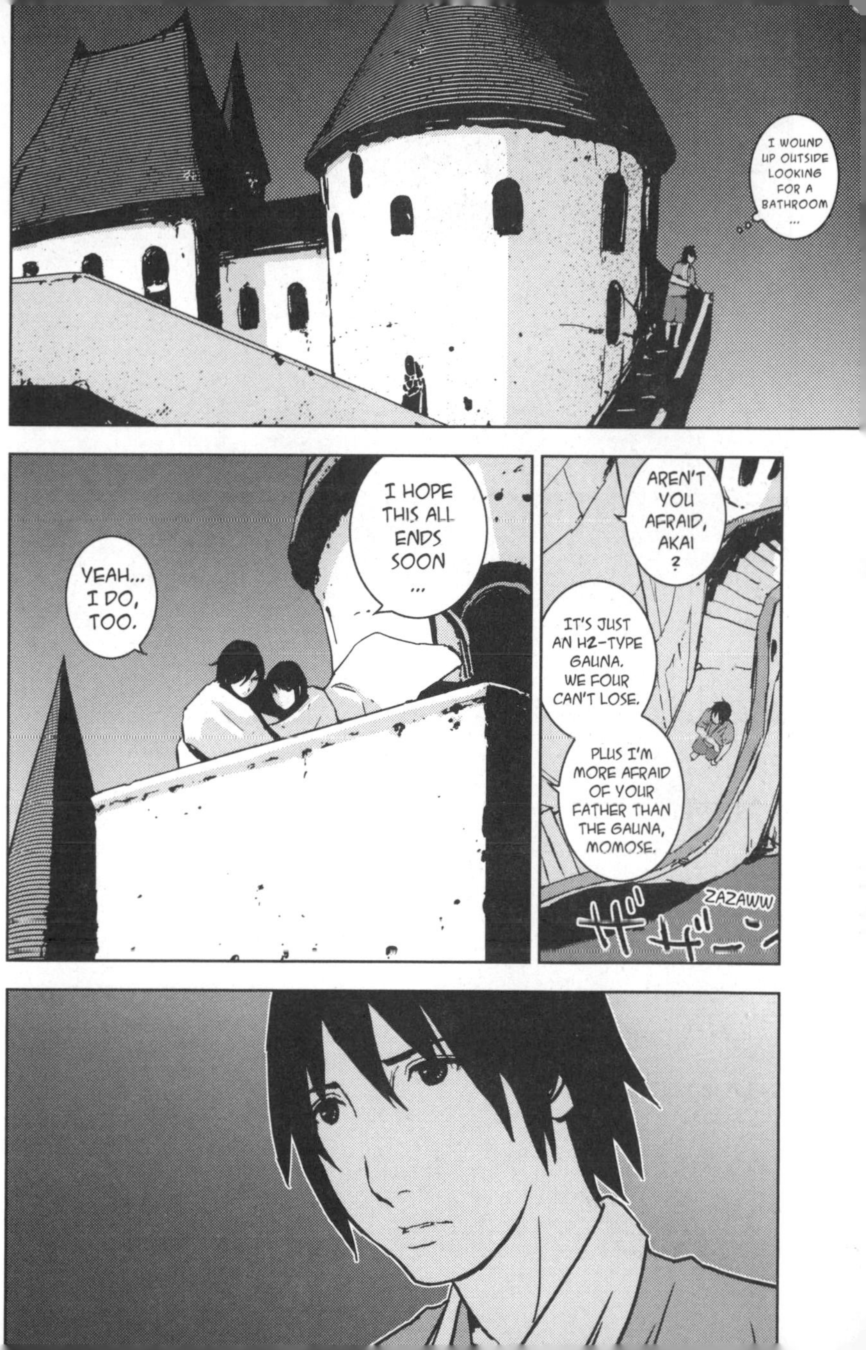
I WOUND UP OUTSIDE LOOKING FOR A BATHROOM ...
AREN'T YOU AFRAID, AKAI ?
IT'S JUST AN H2-TYPE GAUNA. WE FOUR CAN'T LOSE.
PLUS I'M MORE AFRAID OF YOUR FATHER THAN THE GAUNA, MOMOSE.
ZAZAWW
I HOPE THIS ALL ENDS SOON ...
YEAH... I DO, TOO.

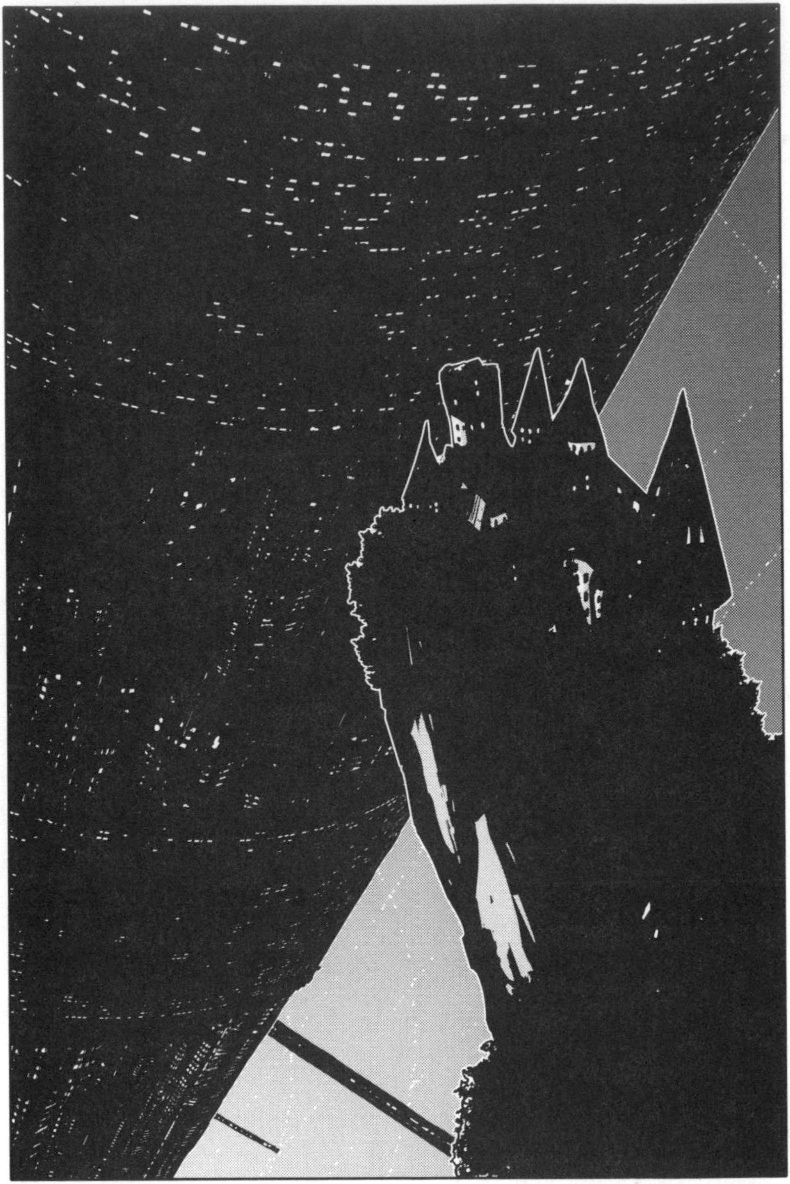

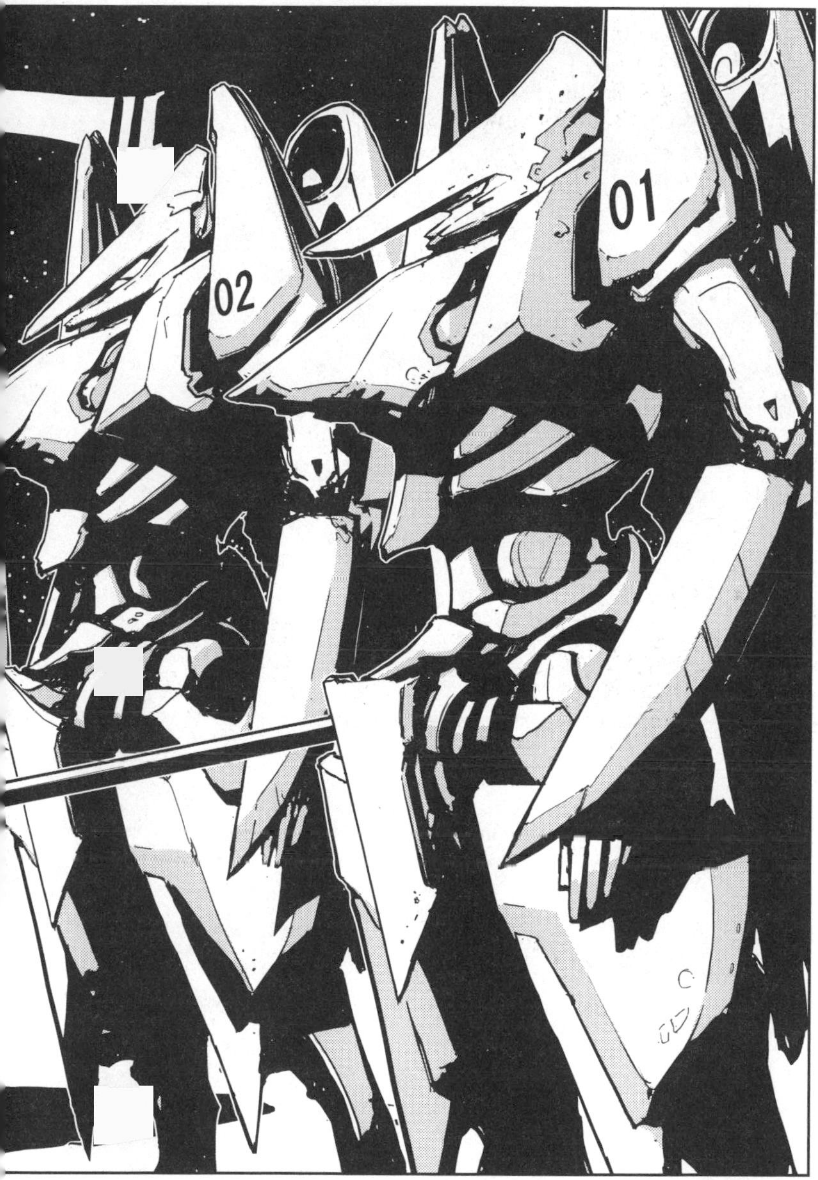
01
02

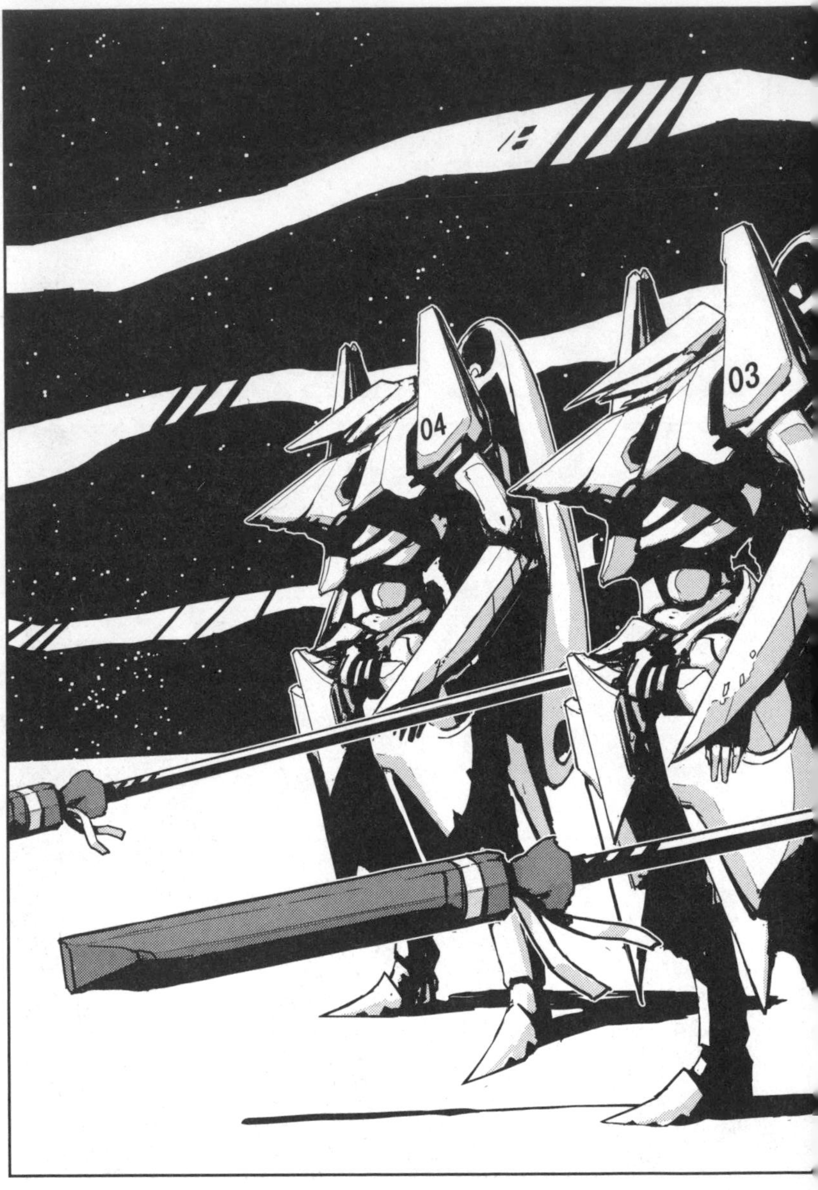
03
04

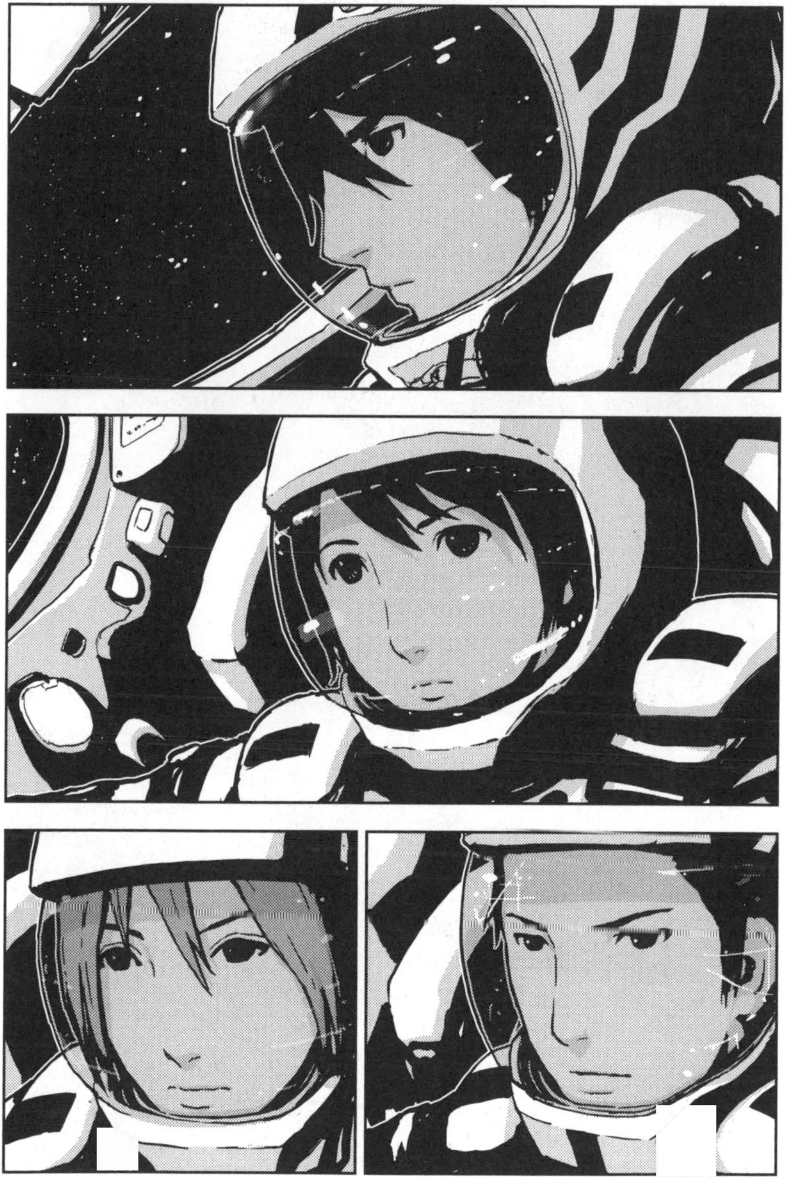

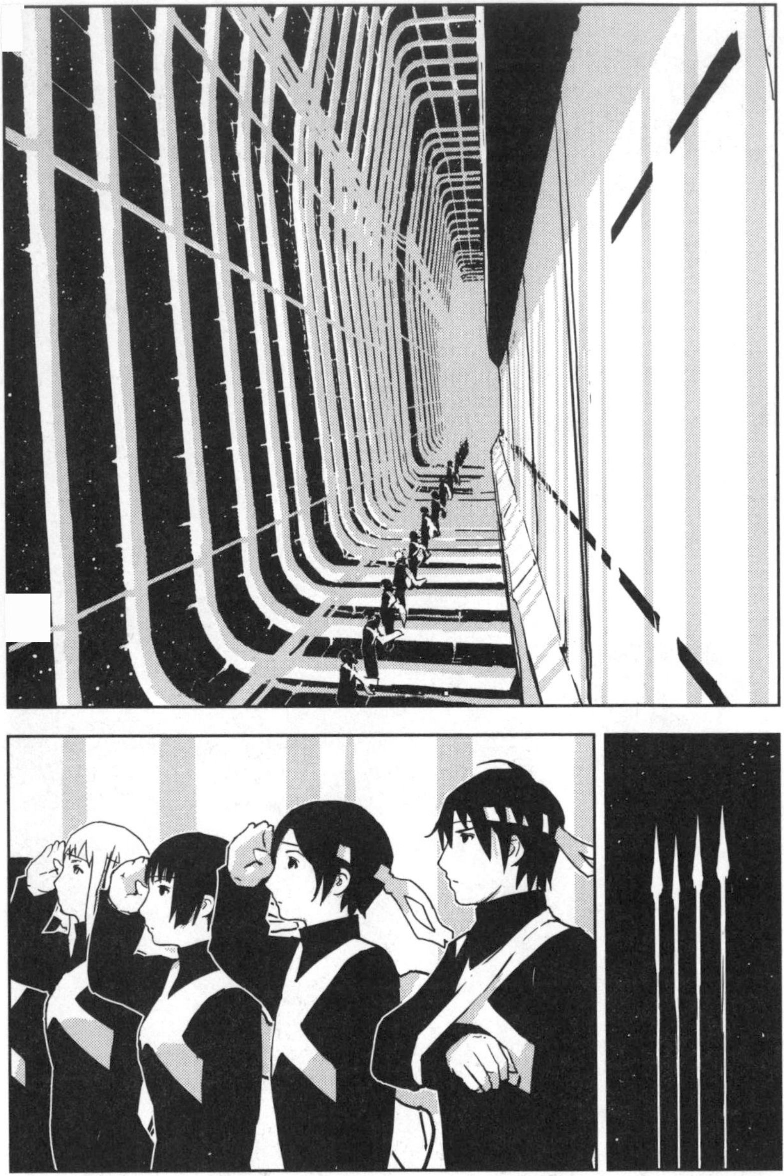

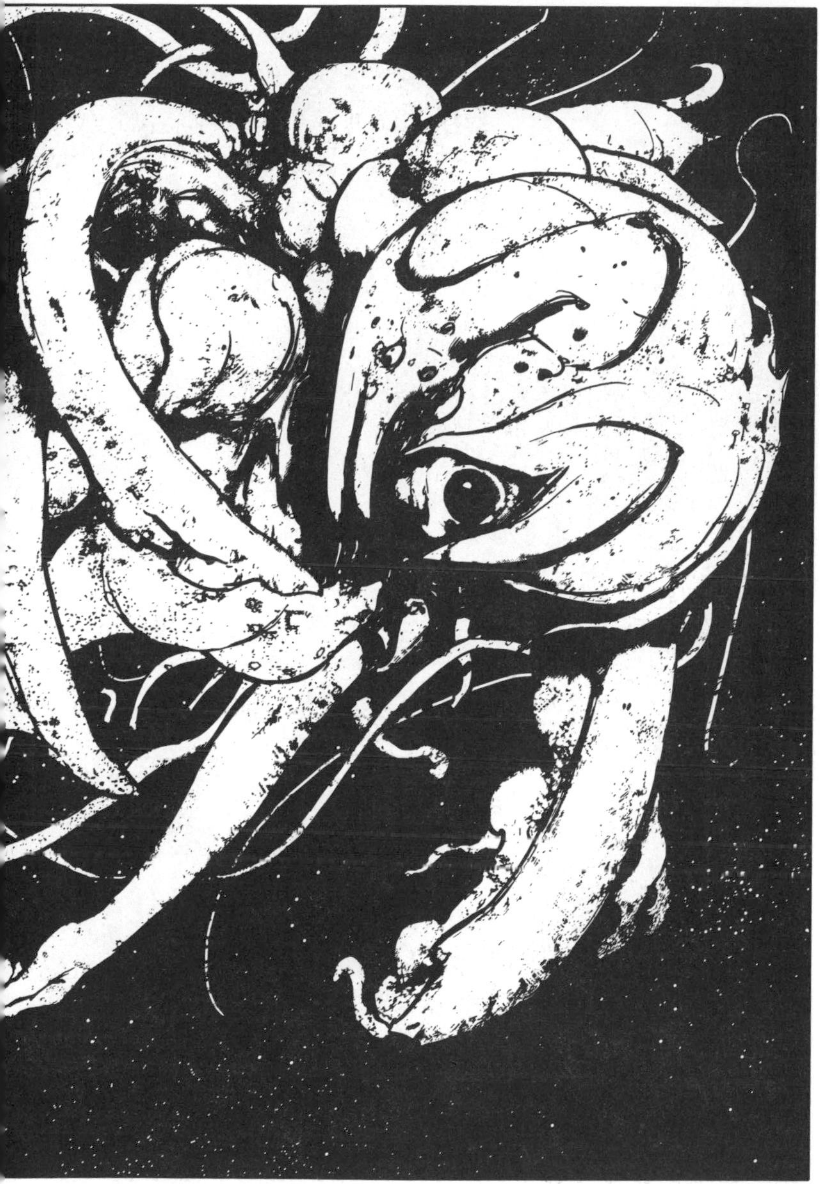

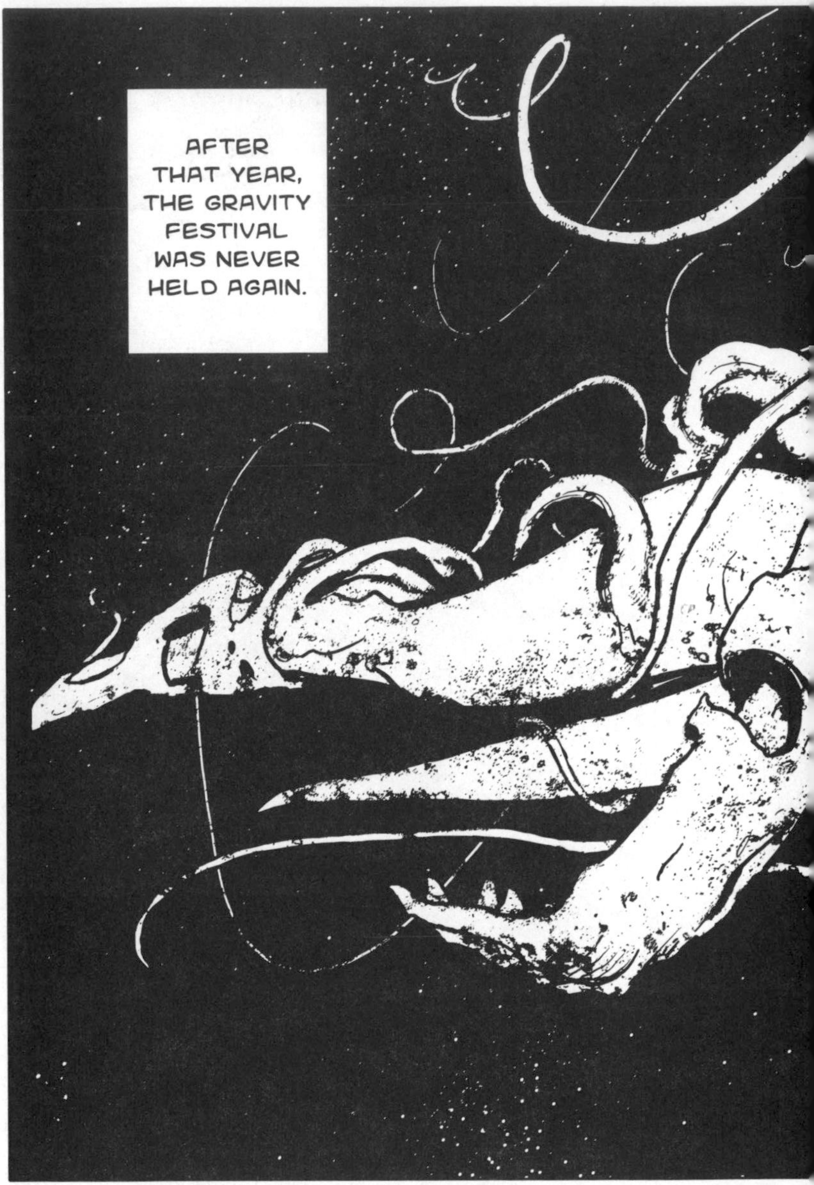
AFTER THAT YEAR, THE GRAVITY FESTIVAL WAS NEVER HELD AGAIN.

KNIGHTS OF SIDONIA Volume ①: END

SOMETHING DEAR,

Nagate trembles as he witnesses the shocking scene before his eyes.

INFINITELY STRONG!!!!

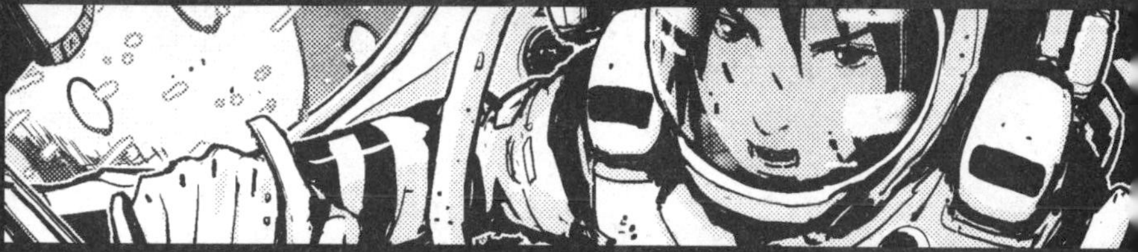

Shocking!!

Moving!!!

to Make the World Shudder!!

IF IT'S TO PROTECT

With the honor and pride of mankind at stake, Sidonia's four best face the Gaur

HUMANS CAN BECOME

Next Volume, Gruesome!

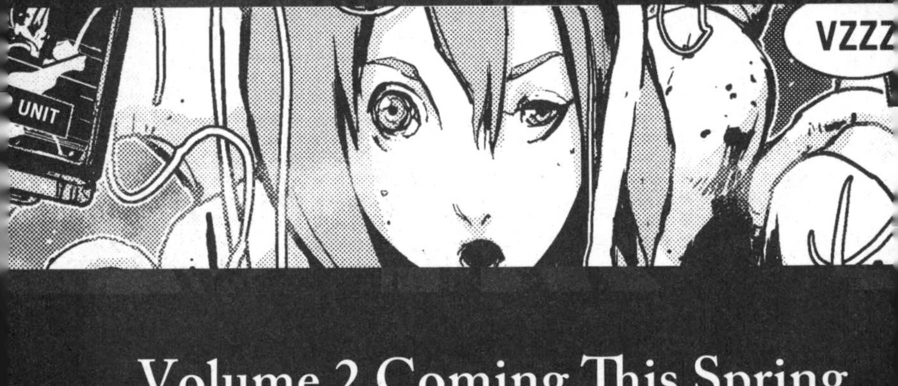
VZZZ
UNIT

Volume 2 Coming This Spring

Knights of Sidonia, volume 1

Translation: Kumar Sivasubramanian
Production: Grace Lu
　　　　　　Nicole Dochych
　　　　　　Daniela Yamada

First published in Japan in 2009 by Kodansha, Ltd., Tokyo
Publication for this English edition arranged through Kodansha, Ltd., Tokyo
English language version produced by Vertical, Inc.

Translation provided by Vertical, Inc., 2013
Published by Vertical, Inc., New York

Originally published in Japanese as *Shidonia no Kishi* by Kodansha, Ltd.
Shidonia no Kishi first serialized in *Bessatsu Shonen Magazine*, Kodansha, Ltd., 2009-

This is a work of fiction.

ISBN: 978-1-935654-80-3

Manufactured in Canada

First Edition

Vertical, Inc.
451 Park Avenue South
7th Floor
New York, NY 10016
www.vertical-inc.com